Expedition Porsche

VNIVERSALE DESCRITTIONE DI TVT
OCEANO SETENTRIONALE
TERRA DE LABORADOR
CIRCVLO ARTICO
TERRA INCOGNITA
NVEVA FRANZA
LARCADIA
GOLFO DI TONZA
OCEANO OCCIDENTALE
TROPICO DE CANCRO
MAR DEL SVR
EQVINOTIALE
PONE NTE
QVITO
BRASIL PRO
CANAS
ATACAMA
CHILACA
CHILI
PLATA
OCEANO MERIDIONALE
TRETO DE MAGALANES
TERRA DEL FVEGO
CIRCOLO ANTARTICO
TERRA INCOGNITA
TERRA MERIDIONALE DISCOPERTO DA NOVO

LA TERRA CONOSCIVTA FIN QVI
OCEANO SYTHICO
MARE MEDITERANEO
MARE DE INDIA
MARE MACADAZO
MARE MACCHIAN
OCEANO ORIENTALE
TROPICHO DE CAPRICORNO
PROVINCIA DE MOLVCO
CAPO DE BONA SPERANZA
TERRA DE LVCACH
TERRA DE VISTA
TER RA IN COGNITA

Vorwort

Zugvögel und Flugmeilensammler müssen jetzt ganz stark sein: Im Jet-Zeitalter mag man zwar innerhalb von Stunden bis ans entgegengesetzte Ende der Welt kommen, aber zum echten Abenteuer wird das banale Absitzen von Flugstunden selten. Außer einem Jetlag ist über den Wolken nichts zu holen, das wahre Fernreisen spielt sich 11.000 Meter weiter unten ab. Am Boden. Wo einem die Landschaften noch unter die Haut gehen, Kilometer für Kilometer.

Honigzäh dahinfließende Ebenen, störrische Gebirge, topografische Umwege als goldene Ideallinie. Die Fortbewegung am Boden lässt deine Seele zu einem hungrig dahintrabenden Wolf werden, der feine Fernweh-Witterung aufgenommen hat und beharrlich vorangeht. Weiter, immer weiter. Irgendwann erreichst du das Ende der dir bekannten Welt, überschreitest die »Weiter-war-ich-noch-nie«-Linie und ab jetzt sickert für lange Zeit eine seltsame Schicksalsergebenheit in die bisher vorherrschende Entschlossenheit. Nun bist du wie auf hoher See unterwegs, navigierst entlang einer aus Unbekannten bestehenden Gleichung, hinein in die weißen Flecken deiner ganz persönlichen, inneren Landkarte. Sicher ist jetzt nur noch das Gestirn der Koordinaten, aber menschliche Gehirne funktionieren eben nicht wie ein Navigationssystem: Himmelsrichtungen, Längen- und Breitengrade sind für uns nur Mittel zur Orientierung, in Wahrheit bewegen wir uns aber entlang von Stimmungen, Erlebnissen und Instinkt.

Reise-Erzählungen spiegeln das wider: Kaum jemand erklärt, wie viele Minuten lang er in welcher Himmelsrichtung unterwegs war und um welche Gradzahl er dann seinen Kurs abgeändert hat, um am Ende diesen oder jenen Punkt auf dem Globus zu erreichen – aber alle Reisenden berichten von der Schönheit des Landes, von den Begegnungen mit seinen Bewohnern, vom Essen und Trinken, von freudigen Überraschungen und zermürbenden Erschwernissen. Gerade deshalb lieben wir die Erzählungen von Fernreisenden so sehr: nicht wegen der schieren Weite, der rein akademischen Anzahl unfassbar vieler Kilometer, sondern wegen dem Mut, der in dieser Art des Reisens steckt. Weil es da ins Unbekannte und Unkalkulierbare hineingeht. Weil Menschen auf weiten Reisen irgendwann ihrer Heimat verloren gehen, die Nabelschnur des Zuhauses reißt und Umkehren länger dauert als der Weg nach vorn.

Foreword

Frequent flyers and those who choose to overwinter in sunnier climes will need to brace themselves: in the jet age, you may be able to reach the opposite end of the world in a matter of hours, but simply racking up airmiles rarely translates into a true adventure. Aside from the problem of jet lag, there's nothing meaningful to be achieved high above the clouds. Real long-distance travel takes place 11,000 meters below. In other words, on the ground, where the landscape can still get under your skin, kilometer after kilometer.

Slow-flowing plains, obstinate mountain ranges and topographical detours become the golden ideal. Traveling overland transforms your soul into a ravenous wolf on the hunt for its next meal, catching the subtle scent of wanderlust and relentlessly pressing onward. Onward, ever onward. Eventually, you reach the edge of your known world, crossing into »unknown territory« and from this point on, your previously dominant determination gives way to a strange resignation to your fate that lasts for a long time. You become like a sailor on the high seas, navigating from one unknown point to the next, venturing into the uncharted spaces of your own personal, internal map. The only certainty now is found in the stars above your head. But human brains don't function like a navigation system: cardinal directions, longitude, and latitude are simply aids to orientation. The truth is that we are guided by mood, experience, and instinct.

Travelogues reflect this fact. Writers rarely recount how many minutes they spent traveling in which direction, or by how many degrees they then altered course to ultimately reach this or that destination. However, all travelers describe the beauty of the country, their encounters with its inhabitants, the food and drink, the joyful highs, and the grueling lows. This is precisely why we love the stories of long-distance travelers so much. We're not interested in the sheer vastness, the purely academic number of unfathomable kilometers, but rather in the courage inherent in this kind of travel. That's because it involves venturing into the unknown and the unpredictable. Inevitably, people making long journeys eventually lose touch with their homeland, the umbilical cord tying them to home snaps, and they reach a point where turning back would take longer than pressing forward. We have the perfect model for the incredibly long journey from the Far East to Europe in the merchant venturer Marco Polo (1254–

Für die ungeheuer weite Reise aus dem fernen Osten nach Europa haben wir mit dem Handelsreisenden Marco Polo (1254–1324) eine perfekte Blaupause, die jahrelangen Abenteuer des Venezianers auf der »Seidenstraße« sind heute noch in unserem Kopf. Dass man auf den über 20.000 Kilometern von einem Ende der europäisch-asiatischen Landmasse zum anderen Ende den ursprünglichen Zweck der Reise aus den Augen verlieren kann, Gefahr läuft nicht wiederzukommen und am Ende vielleicht sogar der Heimat unbekannt geworden zu sein, wissen wir seit Marco Polo. Der Gedanke ist faszinierend. Fährt man vielleicht gerade deshalb so weit? Weil Fernreisen mehr sind als einfach nur eine besonders hohe Entfernung zwischen A und B? Weil man unterwegs verändert wird und nie wieder die Person sein wird, als die man losgefahren ist?

Wir wollten all das selbst erfahren. 15 Gefährten und Reisegenossen, unterwegs in einer Karawane von Porsche-Maschinen. Aus Bangkok, der Hauptstadt Thailands, nach Stuttgart, der Porsche-Hauptstadt ... Die Erlebnisse dieser Reise sollen auf den folgenden Seiten geteilt werden. Weil wir uns die Geschichte selbst noch einmal erzählen müssen, um sie zu glauben. Aber auch, weil wir denken, dass diese Reise ein Privileg ist, das durchs Erzählen geteilt werden kann. Ganz bestimmt aber, weil Reisen Geschichten sind. Mit diesem Buch geht unsere Reise weiter. Kommen Sie mit. Kilometer um Kilometer, soulful driving.

Curves und die Renndrive Crew

1324), whose years-long adventures on the »Silk Road« are still vivid in our minds. Since the days of Marco Polo, we have known the more than 20,000 kilometers from one end of the European-Asian landmass to the other can cause one to lose sight of the original purpose of the journey, risking never returning, and perhaps even becoming a stranger in one's homeland; a fascinating thought. Maybe that's exactly why people travel so far? Perhaps it is because long-distance travel is about more than simply a particularly great distance between A and B? Maybe it's because you are changed along the way and will never again be the person who set out on the journey?

Aiming to find the answers for ourselves, we set out as 15 companions and fellow travelers, taking to the road in a convoy of Porsches. We planned to travel from Bangkok, the capital of Thailand, to Stuttgart, the Porsche capital... Our experiences on this journey will be shared here. This is partly because we need to retell the story to ourselves to be able to believe it, but also partly because we think this journey is a privilege that should be shared through storytelling. Above all, because travel is about stories. This book is the continuation of our journey. Join us on kilometer after kilometer of soulful driving.

Curves and the Renndrive Crew

Der Weg
ist lang,
doch das Herz
kennt keine
Müdigkeit,
wenn ein
Ziel es führt

The path is
long, but
the heart
knows no
weariness
when a goal
leads it

Editorial

Irgendwo zwischen Jiuquan und Ürümqui ist es so weit: Du kommst im Seidenstraßen-Flow an. Ost und West gehen ebenso nahtlos ineinander über wie Vergangenheit und Zukunft, wie das Hinten-Land im Rückspiegel und das Vorne-Land vor der Windschutzscheibe. Zwar zeigt der zielsichere Pfeil auf dem Navigations-Display genauso entschlossen geradeaus wie seit Stunden, aber du löst dich von seinem Diktat. Zählst nicht mehr die Kilometer, denn die haben jede Bedeutung verloren, sondern gehst im Moment auf: Vor und hinter dir zieht gerade ein staubiges Asphaltband durch monochrome Weite. In der Ferne ragt eine Gebirgskette empor, die seit Stunden am selben Fleck zu stehen scheint. Die einzige Abwechslung besteht aus dem immer wiederkehrenden Rhythmus von Raststellen, die sich mit ihren ausgeblichenen Spritpreis-Schildern und Hinweisen auf morbide Proviant-Optionen als neben die Straße geduckte Flachbauten manifestieren. Kleine Gruppen von Kamelen haben daneben die Köpfe in den Steppenwind gedreht, ihre zottigen Mähnen zappeln in der unaufhörlichen Brise. – Und während du dieses starke Bild in dich aufnimmst, weißt du, dass es dich nie wieder verlassen wird. Eingebrannt in der Erinnerung.

Dann melden sich die Fata-Morgana-Gedanken des Fernreisens: »Die Welt könnte ja auch ein Laufband sein,« denkst du, »das sich unter uns dreht, und wir stehen in Wahrheit still, ganz still.« – Oder diese andere, skurrile Vorstellung: »Man könnte doch bestimmt den Mekong, den Jangtsekiang oder Huáng Hé in den heißen Sand der Taklamakan lenken, und selbst von deren Wassermassen würde irgendwann nur ein letzter Tropfen übrigbleiben, bevor auch der zischend vergeht«. Bittersüße Wahrheit. Selbst die Größten, Mächtigsten finden immer ihren Endgegner – und wenn es nur die Zeit ist, die pure Vergänglichkeit ... Hier draußen wird das zur zwingenden Metapher. Du fährst staunend, ganz ergriffen von der immensen Weite der Welt. Von Curvistan nach Absurdistan. Und genau in diesem Moment zerreißt es dich innerlich beinahe vor Freude und Dankbarkeit, absoluter Neugierde und ungeheurer Abenteuerlust.

Du gehörst zu den ganz Wenigen, die je das Privileg hatten, die unfassbare Weite Asiens und Europas zu durchqueren, die schier unglaubliche Freiheit einer solchen Fahrt zu erleben. Ein Strudel von Bildern und Eindrücken hat sich in Deine Seele gelegt. Unverarbeitet. RAW-Format. Gut, dass menschliche Gehirne

Editorial

Somewhere between Jiuquan and Ürümqui, you finally get there: you find yourself in the flow of the Silk Road. East and West merge seamlessly, as do past and future. The land you've left behind is a wan reflection in your rearview mirror while the land ahead looms before your windshield. The remorseless arrow on the navigation display still points straight ahead, just as it has for hours, but you are ready to break free from its diktats. You stop counting kilometers, for they have lost all meaning, but instead become absorbed in the moment: a dusty ribbon of asphalt stretches fore and aft through a monochrome expanse. In the distance, a mountain range rises, seemingly frozen in place for hours. Variation is only afforded by the recurring rhythm of rest stops, low-rise buildings crouching beside the road with faded fuel price signs and baleful hints at food offerings. Small groups of camels huddle nearby, their heads turned toward the wind of the steppes, their shaggy manes fluttering in the ceaseless breeze. As you absorb this powerful image, you know it will never leave you. Seared into your memory.

It is at this point that the delusional thoughts of long-distance travel start to surface: »The whole world could be a like a giant conveyor belt,« you think, »turning beneath us, and we're actually standing completely still.« Or perhaps a different bizarre notion: »Surely it would be possible to divert the Mekong, Yangtze, or Huang Hé to the hot sands of the Taklamakan Desert. Even then, every last drop of their vast waters would eventually vanish with a hiss.« There's bittersweet truth at play here. Even the greatest and most powerful forces eventually meet their ultimate adversary – even if that adversary is just Old Father Time himself, in other words simple transience... Out here, this thought becomes a compelling metaphor. You drive in awe, utterly captivated by the immensity of the world as you wend your way from Curvistan to Absurdistan. And at that very moment, you're almost bursting with joy and gratitude, absolute curiosity, and an overwhelming thirst for adventure.

You are among the very few who have ever had the privilege of crossing the unfathomable vastness of Asia and Europe and experiencing the sheer incredible freedom of such a journey. A whirlwind of images and impressions has washed over your soul unprocessed and in raw format. Fortunately, although human

Die Glocken und Klanghölzer der Tempel Südostasiens, das hektische Huporchester des Verkehrs in den Millionenstädten, der erste, vorsichtige Biss in einen frisch gegrillten Fleischspieß am Straßenrand. Bunt wirbelnde Gewänder, urbane Soundkulissen, das Plätschern von Regen in den Blättern der Bäume oder das stetige Rauschen des Windes, draußen in der Steppe. – Immer wenn du dich daran erinnerst, folgt die ganze Szene in aller Opulenz dicht auf den Fuß, als würde deine Erinnerung ihre Bilder überlebensgroß hinausschicken, intensiv und laut wie eine Marching Band beim Karneval von New Orleans.

zwar keine unbegrenzte Speicherkapazität haben, aber im Reise-Modus die Fähigkeit zu sinnvoller Komprimierung: Geschichten werden zusammengesetzt, Verbindungen geknüpft, Erlebnisse und Eindrücke in die passende Reihenfolge gewoben. Als Zugangscodes in die abgespeicherten Erinnerungen dienen später dann die dazu abgelegten Düfte und Geschmäcker, Musik und Klang: Die Glocken und Klanghölzer der Tempel Südostasiens, das hektische Huporchester des Verkehrs in den Millionenstädten, der erste, vorsichtige Biss in einen frisch gegrillten Fleischspieß am Straßenrand. Bunt wirbelnde Gewänder, urbane Soundkulissen, das Plätschern von Regen in den Blättern der Bäume oder das stetige Rauschen des Windes, draußen in der Steppe. – Immer wenn du dich daran erinnerst, folgt die ganze Szene in aller Opulenz dicht auf den Fuß, als würde deine Erinnerung ihre Bilder überlebensgroß hinausschicken, intensiv und laut wie eine Marching Band beim Karneval von New Orleans.
Aus der Zartheit subtilster Augenblicke, aber auch dem Schönheits-Schock überraschender Groß-Sensationen hat dein Unterbewusstes ein buntes Bild geschaffen. Während es von außen wirkt, als seist du auf Autopilot im Meilenfresser-Wachkoma unterwegs, arbeitet dein Gehirn auf Hochtouren, pegelt derweil deinen Glückshormonspiegel auf Dauermaximum ein. »Das hier,« denkst du grinsend bei Kilometer 7.236 oder 10.484 oder während du aus einer der letzten Kolonnen-Positionen zusiehst, wie die Autos deiner Reisegefährten vor dir vorsichtig um ein besonders großes Schlagloch kurven, »ist das Beste, was ich

brains don't have unlimited storage capacity, they do possess the ability to meaningfully compress information when on the move: stories are pieced together, connections are made, and experiences and impressions are woven into the correct sequence. The scents and tastes, music and sounds stored in these memories later serve as keys to unlock further reminiscences: the bells and wood blocks of the temples of Southeast Asia, the horn orchestra of hectic traffic in the megacities, the first, cautious bite into a freshly grilled meat skewer purchased on the roadside. Colorfully swirling garments, urban soundscapes, the patter of rain on the leaves of the trees, or the constant rushing of the wind out in the steppes. Whenever you recall it, the whole scene follows in all its opulence, as if your memory were projecting a set of larger-than-life images, intense and overwhelmingly loud like a marching band at the New Orleans Mardi Gras.

Your subconscious has assembled a vibrant collage from the tenderness of the tiniest moments and the shock of surprising, grand sensations. While to an outsider it may seem like you're cruising along in a relaxed fashion, gobbling up the miles as if on autopilot, your brain is in fact working at full speed, all the while maintaining a constant peak in your endorphin levels. »This is the best thing I've ever experienced,« you think with a grin at kilometer 7,236 or 10,484, or as you watch from the tail end of the convoy as your fellow travelers carefully navigate around a particularly large pothole.

So, how did it all begin? As so often, with a dream, a passion. In our case, it all started with the age-old fascination with being on the road. It's been part of us for as long as we can remember. But it's not just an insatiable curiosity about the land just beyond the next bend that drives us forward, but rather the bends themselves. Each one is a call to action, to change direction, to stimulate new neural links, and to gain new insights. People who sit still remain stagnant. And that's why we dream about the world beyond, traveling to lose ourselves and then find ourselves again. Maybe that's how we define wanderlust.

Why did we choose Bangkok of all places as our starting point? Is it because it's far enough away? Or maybe simply because you have to start somewhere? Is it because that's where most of our small, close-knit group of travelers comes from? – To be honest,

je erlebt habe!« Wie alles begonnen hat? Wie so oft: mit einem Traum, einer Leidenschaft. In unserem Fall mit der uralten Faszination für das Unterwegssein. Sie steckt in uns, seit wir denken können. Nicht nur eine unersättliche Neugierde auf das Land hinter der nächsten Kurve treibt uns voran, sondern sogar die Kurven selbst. Jede einzelne ist für uns ein Call-to-Action, zur Richtungsänderung, zum Synapsen-wachsen-Lassen und zum Erkenntnisgewinn. Wer sitzt, bleibt stehen. Und deshalb träumen wir von der Welt, sind unterwegs, um verlorenzugehen und uns genau darin wiederzufinden. Vermutlich ist das unsere Definition von Fernweh.

Weshalb wir ausgerechnet in Bangkok losgefahren sind? Vielleicht, weil das weit genug weg ist? Oder weil man irgendwo ja losfahren muss? Weil da die Mehrheit unserer kleinen, eingeschworenen Reisegruppe herkommt? – Nun denn: von allem ein wenig. Vermutlich hat jeder in unserer eingeschworenen Reisegesellschaft ein ganz eigenes Ziel, das sogar er selbst nur erahnen kann, aber die offizielle Begründung geht so: Wir sind Freunde. Mit einer gemeinsamen Faszination fürs Autofahren, die wir als »Renndrive Crew« ausleben.

Nicht selten zieht es die Renndrive Crew aus Bangkok hinauf in den Norden Thailands, bis nach Bo Kluea und in den Doi-Phu-Kha-Nationalpark, dort sind die Straßen leergefegt und teilweise trotzdem exzellent asphaltiert. Diese 300-Kilometer-Runde ist sozusagen die Hausstrecke der Renndrive Crew und irgendwann kam, was kommen musste: Beim Umkehren am fernsten Punkt haben wir die Straße nach Norden hinuntergeschaut und gedacht: »Was, wenn man einfach weiter fahren würde ...? Bis zur Grenze nach Laos am Mekong, weiter durchs grüne Nachbarland im Norden? Exakt 222 Kilometer später hätte man die Grenze nach China erreicht und dann wäre es bereits zu spät umzukehren. Man müsste in zwingender Logik weiterfahren ...« – Obwohl unser Freundeskreis »Renndrive Crew« heißt und nicht »Cruise Crew«, hat sich diese Flause erstaunlich schnell festgesetzt. Wir finden das übrigens vollkommen logisch: Fahrer, für die das »soulful driving« zum Lebenselixier gehört, neigen schnell zu besonders weiten Fahrten. Man will einfach nicht aufhören, nicht auschecken aus dem Flow, den Rausch nicht unterbrechen. Die 24-Stunden-Rennen im Motorsport, die Mille-Miglia-Races, Panamericana-Rallyes, Paris-Dakar-Abenteuer

a little bit of all of those things. Presumably, every member of our tight group has a personal goal of their own, one that even they themselves can only vaguely guess at. However the official explanation goes something like this: we're friends who share a fascination for driving, which we indulge as the self-styled »Renndrive Crew« (»Racing Crew«).

The Renndrive Crew often heads from Bangkok to northern Thailand, all the way to Bo Kluea and Doi Phu Kha National Park. There, the roads are practically deserted, yet some of them have excellent surfaces. This 300-kilometer loop is the Renndrive Crew's home turf, so to speak, so of course the inevitable eventually happened: turning around at the furthest point, we looked along the road to the north and thought, »What if we just kept going...? All the way to the border with Laos on the Mekong, continuing through the verdant neighboring country to the north? Exactly 222 kilometers later, we would reach the border with China, and by then it would already be too late to turn back. Logically, we would have to keep going...« – Although our group of friends is known as the »Renndrive Crew« rather than the »Cruising Crew,« this idea took hold surprisingly quickly. We view this as perfectly logical, by the way: Drivers for whom »soulful driving« is an essential part of their lives tend to be particularly fond of long journeys.

They simply don't want to stop because to drop out of the flow would risk interrupting that euphoric feeling. The classic 24-hour races in motorsport, the Mille Miglia races, the Panamericana rallies, the Paris-Dakar adventures, and the Targa Florio fever dreams all stem from the same mindset: driving for the love of driving. There is a reason why these Gran Turismo races or multi-day rallies are driven against the clock – thus setting a predefined end to the pleasure, rather than waiting for someone to breakdown in tears, die, run out of fuel, or write off their car (after all, these things can happen...). That reason is probably to be found in humanity's delight in meaning, significance, and rituals: our lives are finite, so we tend to set limits, even in ecstatic moments. It's no coincidence that we measure cars drily in »kilometers per hour« instead of »excitement per experience.« That's also why we gave our journey a concrete beginning and end and defined stages in between: »All we needed to do was simply drive the Porsche all the way home to Stuttgart, so to speak.«

und Targa-Florio-Fieberattacken entstammen allesamt demselben Mindset: Fahren um des Fahrens willen. Der Grund, weshalb man diese Gran-Turismo-Rennen oder Mehrtage-Rallyes gegen die Uhr fährt – also dem Vergnügen eben doch ein vordefiniertes Ende setzt – und nicht, bis jemand weint, stirbt, der Sprit ausgeht, die Maschine zerbricht (na gut, das alles ist gar nicht so selten ...), ist vermutlich in der Freude des Menschen an Sinn und Bedeutung und Ritualen zu suchen: Unsere Lebenszeit ist endlich, also neigen wir dazu, sie selbst in den ekstatischen Momenten zu begrenzen. Nicht umsonst messen wir Autos in der trockenen Einheit »Kilometer pro Stunde« statt in »Excitement per Experience«. Und deshalb haben auch wir unserer Fahrt einen konkreten Beginn, ein konkretes Ende und definierte Etappen dazwischen gesetzt: »Man müsste mit den Porsche einfach mal bis nach Stuttgart fahren, nach Hause sozusagen.« Naheliegend? – Na klar! Wir sind Porsche-Fans, und das hier ist unsere Pilgerfahrt. Bangkok to Stuttgart, the epic drive – das wurde zum Arbeitstitel unserer Reise ins Zuhause des Neunelf und ins Zentrum des Porsche-Universums: Wir werden ganz vom Rand kommend hineinfahren und dann, wer weiß, bunte Blumenkränze ablegen. Wenn es die denn in Stuttgart-Zuffenhausen beim Blumenladen um die Ecke gibt. Die aus Bangkok wären nach 22.000 Kilometern bereits etwas angegriffen ...

Wir, das sind – Tenn, Chayanin, Oou, Humphrey, Suthi, A GT, Lit, Art, Stefan, Jack, Aob, Mog, Joke, Gap und Ake. Tenn heißt eigentlich Sihabutr Xoomsai, der Filmregisseur ist Vordenker der »Renndrive Crew« und Gründer des Porsche-Fests »Das Treffen« in Bangkok. Porsche-Fans aus ganz Südostasien kommen hierher, Anziehungskraft und Vibe von »Das Treffen« sind bemerkenswert. Tenn fährt ganz unromantisch seinen 2024er Porsche Cayenne S E-Hybrid, bietet sich damit als Navigator sowie Lastenträger für Survival-, Film- und Foto-Equipment an. Auf dem Beifahrersitz des schnellen Reise-Mobils sitzt Stefan, der Münchner Fotograf und Medienmacher steckt hinter dem Magazin »CURVES«, ist ein Kristallisationspunkt der Porsche-Szene, Alpen-Insider, Reise-Tier. Stefan wird im Laufe der Fahrt als einziger Europäer der Reisegesellschaft seiner Heimat immer näher kommen, anstatt sich von ihr zu entfernen. Er wird den epischen Ritt dokumentieren – und auch mal für ein paar hundert Kilometer ins Lenkrad fassen, wenn den Kollegen die Hände zittern.

Is this all a bit obvious? – Of course! We're Porsche fans after all, and this is our pilgrimage. Bangkok to Stuttgart, the epic drive became the working title for our journey to the home of the 911 and the heart of the Porsche universe: we would arrive from the margins and then, who knows, perhaps lay down a carpet of colorful flower garlands. That is assuming that they sell them at the florist on the corner in Stuttgart-Zuffenhausen. Flowers brought all the way from Bangkok would be a bit tattered after 22,000 kilometers...

The crew comprises Tenn, Chayanin, Oou, Humphrey, Suthi, A GT, Lit, Art, Stefan, Jack, Aob, Mog, Joke, Gap and Ake. Tenn's real name is Sihabutr Xoomsai. A film director, he is the mastermind behind the »Renndrive Crew« and founder of the »Das Treffen« Porsche festival in Bangkok. Porsche enthusiasts from all over Southeast Asia flock to the event for its remarkably appealing vibe. Eschewing romance, Tenn drives his 2024 Porsche Cayenne S E-Hybrid, offering to act as navigator and cargo carrier for survival equipment as well as our film, and photographic gear. The passenger seat of the speedy roadster is occupied by Stefan, the Munich-based photographer and media producer behind »CURVES« magazine, a key figure in the Porsche scene and a true traveler, with unrivaled insider knowledge of the Alps. As the journey progresses, Stefan, the only European in the group, will find himself drawn ever closer to his homeland rather than travelling in the opposite direction as is his custom. He will document the epic ride, even taking the wheel for a few hundred kilometers when his fellow crew members' hands have had enough.

Right behind the lead car, things start to get more adventurous – Chayanin has decided to drive his converted »Safari« all the way to Germany. Beneath the yellow paint and racing livery beats the heart of an air-cooled 911, built in 1986. The question of whether this rugged youngtimer will reach our destination in Stuttgart is the subject of several bets. »Oou« is keeping pace in a Porsche 987 Cayman S. This compact mid-engined sports car has also undergone a modest off-road makeover with wider track and significantly increased ground clearance for the more challenging sections of the Asian outback.Humphrey, on the other hand, is far less concerned about the challenges that may be posed by the road surfaces (or lack thereof...) between Thailand

Direkt dahinter wird es abenteuerlich – Chayanin hat beschlossen, es in seinem »Safari«-Umbau bis nach Deutschland zu schaffen. Unter dem gelben Lack und der Racing-Kriegsbemalung steckt ein luftgekühlter 911, Baujahr 1986. Ob der ruppige Youngtimer das Ziel in Stuttgart sehen wird, ist Gegenstand von mehreren Wetten. »Oou« hält im Porsche 987 Cayman S mit, auch der kompakte Mittelmotorsportler hat eine milde Offroad-Kur hinter sich: Spurverbreiterung und konsequent angehobene Bodenfreiheit, für die nicht ganz jugendfreien Passagen im Outback Asiens.

Deutlich sorgloser, was die Herausforderungen des Straßenbaus (oder Nicht-Baus ...) zwischen Thailand und dem fernen Europa angeht, ist dagegen Humphrey unterwegs: Der vipergrüne 992 GTS will es mit serienmäßiger Bodenfreiheit quer durch den Wilden Osten schaffen. Immerhin ist das Auto ein Exemplar der limitierten »Thailand Edition« und diese Vorschusslorbeeren sollten durch sämtliche Widrigkeiten helfen. Findet Humphrey zumindest, der Rest der Reisegesellschaft hat an seine Wette aufs Ankommen erst im Moment der Zieldurchfahrt geglaubt, ihm das aber 22.000 Kilometer lang verschwiegen ... Und dann wären da noch die Autos, um die man sich wirklich überhaupt keine Sorgen macht. Für den von Suthi pilotierten Porsche 992 Dakar ist diese Reise wie gemacht, und das Begleit-Team setzt gleich auf vollkommen Komplikations-unverdächtiges Material: Jack darf im Mercedes-G-Modell dabei sein, weil der quasi die Ikone der anderen Stuttgarter Automarke ist, sozusagen Gegenpol zum Porsche 911. Und dass die »Support Crew« der »Renndrive Crew« mit Joke, Mog und Gap im Toyota Landcruiser unterwegs ist, darf als demütige Verbeugung vor universell gültigen Fernreise-Standards betrachtet werden: Wenn du einen Landcruiser im Rückspiegel hast, kann alles einfach nur gut enden. Und da wäre noch Ake, unser Mastermind der die Reise bis ins Detail geplant – er ist die Tour schon mehrfach gefahren und plant nach unserer Ankunft in Stuttgart gleich weiter ans Nordkap weiterzureisen.

In dieser Gelassenheit sind wir den ganzen, weiten Weg gefahren, haben der Zeit als ultimativem Endgegner die enormen Distanzen mit schierer Hartnäckigkeit und tiefer, satter Freude entwunden. Und an alle Zuhausegebliebenen senden wir nun eine gute Nachricht: Diese Welt ist wunderschön, im ganz Großen

Und an alle Zuhausegebliebenen senden wir nun eine gute Nachricht: Diese Welt ist wunderschön, im ganz Großen und winzig Kleinen. Man kann fühlen und schmecken, hören und erschnuppern, wie sich die Kulturen aneinanderschmiegen, jede an ihren Nachbarn, wie die Gemeinsamkeiten und Unterschiede von Ost bis West ineinanderfließen. Überall ist Lachen und lebhaftes Interesse, das angebliche Gegeneinander der Menschheit haben wir auf 22.000 Kilometern nie gesehen.

We have good news for all those who stayed home: this world is truly beautiful, in the grandest of ways and in the smallest of details. You can feel and taste, hear and smell how cultures bump up against one another, how the similarities and differences merge together as you move from East to West. Everywhere we are greeted with laughter and lively interest. We never once witnessed the supposed antagonism between humanity over 22,000 kilometers.

---

and faraway Europe. Featuring standard ground clearance, his viper-green 992 GTS is determined to make it across the wilds of the East. After all, the car is a limited »Thailand Edition«, an initial accolade that should help it overcome any adversity. At least, that's what Humphrey believes. The rest of his traveling companions refused to believe that he would reach the finish line until the moment he crossed it, something they kept secret from him for 22,000 kilometers... Then there are the cars that no one worries about at all. This trip is tailor-made for Suthi's Porsche 992 Dakar, and the support team has opted for guaranteed trouble-free equipment: Jack is allowed to join the group in a Mercedes G-Class, because it is more or less the mascot of the other Stuttgart-based car brand, almost the antithesis of the Porsche 911. The fact that the »Renndrive Crew« support team, Joke and Moh, are driving a Toyota Land Cruiser can be interpreted as a humble nod to universally accepted long-distance travel standards: If you have a Land Cruiser in your rearview mirror, you know that all will be well. And then there's Ake, our mastermind who planned the journey down to the smallest detail. He has already driven the tour several times and, after our arrival in Stuttgart, is planning to continue straight on to

und winzig Kleinen. Man kann fühlen und schmecken, hören und erschnuppern, wie sich die Kulturen aneinanderschmiegen, jede an ihren Nachbarn, wie die Gemeinsamkeiten und Unterschiede von Ost bis West ineinanderfließen. Überall ist Lachen und lebhaftes Interesse, das angebliche Gegeneinander der Menschheit haben wir auf 22.000 Kilometern nie gesehen. Wir glauben es nur, weil wir einander nicht in die Augen sehen. Und genau das war auf der Pilgerfahrt von Ost nach West nie möglich: In jedem Dorf, an jeder Tankstelle, auf jeder Autobahn-Etappe haben wir nur Begeisterung für unsere Reise erlebt. Sobald die Menschen verstanden haben, woher wir kommen und wohin wir fahren: sprühende Faszination! Ausgelassenes Staunen! Absolute Sympathie!

Und wir haben eine Welt erlebt, die wir in unseren kühnsten Träumen nicht vermutet hätten. Manchmal sind wir in unserer Erinnerung immer noch auf den Dschungelstraßen von Laos unterwegs, rau und verwegen und wild. Dann erinnern wir uns an die Grenze nach China, an die Macher-Atmosphäre dieses Landes, an das explosive Wachstum, die turbulenten und kunterbunten, weit in den Himmel ragenden Städte. An glitzernde Wolkenkratzer-Skylines in den schillernden Nächten von Millionenstädten, deren Namen wir nie zuvor gehört haben. Aber jetzt waren wir dort: in Kunming, Qujing, Xi'an und Lanzhou. Und die Namen sagen uns nun etwas, sie sind wie lebendige Gesichter geworden. Wir haben Augenkontakt aufgenommen. Wirklich gesehen, mit den äußeren Augen und den inneren. Genau wie Kasachstan, dieses ungeheure Land mit seinen Wüsten, Steppen und Gebirgen. Hier ist die Natur ein Titan, und Menschen sind vergängliche Zwerge. Kasachstan steckt jetzt unter unserer Haut, als immer wieder abrufbares und kristallines Fernweh. Dass aus diesem Raum Asiens heraus Menschheitsgeschichte geschrieben wurde, haben wir vor unserer Reise theoretisch verstanden – aber jetzt spüren wir es.

Hinter dem Schwarzen Meer ist die Welt eine andere: Das wilde Georgien geht in den asiatischen Teil der Türkei über, und plötzlich tauchen uralte Geschichten auf. Für Reisende aus Südostasien ist die Wiege der Menschheit an der Kreuzung zwischen Afrika, Europa und Asien weit entfernt. Auf einmal hier unterwegs zu sein, in den Ländern der Perser, Skythen, Hethiter, Griechen ... und die Geschichte mit all ihren Facetten und Mysterien immer noch zu erahnen, ist ein ausgesprochen anrührendes Erleb-

the North Cape We covered the entire, long trip in this serene mood, applying sheer tenacity and deep joy and satisfaction to wrestle the enormous distances from the grasp of time, the ultimate adversary. We have good news for all those who stayed home: this world is truly beautiful, in the grandest of ways and in the smallest of details. You can feel and taste, hear and smell how cultures bump up against one another, how the similarities and differences merge together as you move from East to West. Everywhere we are greeted with laughter and lively interest. We never once witnessed the supposed antagonism between humanity over 22,000 kilometers. Perhaps we only believed this because we were never able to look each other in the eye on our pilgrimage from East to West: in every village, at every gas station, on every stretch of highway, we encountered only enthusiasm for our journey. As soon as people understood where we came from and where we were going, we encountered bubbling fascination, unbridled awe, and absolute empathy.

We experienced a world we wouldn't have dared to imagine in our wildest dreams. Sometimes, our minds linger on the rough, rugged, and wild jungle roads of Laos. Then we remember crossing the border into China, the can-do atmosphere of that country, its explosive growth, the vibrant and colorful cities that reach high into the sky. Glittering high-rise skylines in the dazzling nightlights of megacities we'd never heard of before. But now there we were in Kunming, Qing, Xi'an, and Lanzhou. These names mean something to us now, taking on a living character. We made real contact, seeing these places with our own eyes and absorbing the sights, sounds and smells deep into our souls. It was the same with Kazakhstan, an immense country of deserts, steppes, and mountains. Here, nature is gargantuan, while humans are like ephemeral specks of dust. Kazakhstan really got under our skin, awakening a constantly accessible, sharp longing for distant lands. Before our trip we understood in theory that human history was written in this region of Asia, but now we have come to feel it. The world is quite different beyond the Black Sea. The wilds of Georgia merge into the Asian part of Türkiye, and suddenly ancient stories emerge. For travelers from Southeast Asia, the cradle of humanity at the point where Africa, Europe, and Asia meet seems a world away. It is a profoundly moving experience suddenly to find yourself here, in the lands of the Persians, Scythians, Hittites and Greeks and still be able to sense history with all its facets and mysteries.

nis. Und dann bist du auf einmal in Europa. Eine Brücke in der Türkei führt aus Asien hinüber, es geht schockierend schnell. Für die letzte Etappe muss man sich dann schon etwas Zeit lassen, verlässt besser den direkten Kurs und schleicht sich durch die Hintertür zum Ziel. Durch den kernigen Norden Griechenlands und die wilden Gegenden Bulgariens und Rumäniens, mit einem weiten Bogen über Ungarn zurück in die Balkan-Länder. Europäer sagen, erst hier würde man dann spüren, wie aus dem Osten der Westen wird, eine ganz andere Facette der Menschheit. Aber man kann die Welten verbinden, ganz leicht. Muss einfach nur: fahren.

Oh, und für Alle, die von der weiten Reise zwischen Asien und Europa aus den verschiedensten Gründen immer nur träumen können, haben wir eine Empfehlung – es gibt eine alternative Route. Sie beginnt in den herrlich aromatischen Garküchen Südostasiens und legt dann in den unglaublich diversen Küchen der Regionen Chinas einen ersten Stopp ein: China ist ein kulinarischer Kontinent. Für die weitere Reise sollte man Fleisch mögen – Vegetarier oder gar Veganer haben es in den Weiten Kasachstans schwer, und diese Küchen-Grundstimmung schafft es – nach Lammfett-duftend und Grill-Aroma-schwanger – bis weit in den Westen. Bis auf den Balkan und nach Griechenland reicht diese Grillspieß-Kultur, hier schleichen sich dann die Gemüse-Sorten der alten und neuen Welt ein. Aus Italien herauf schiebt eine kulinarische Supermacht ihre Gravitationsenergie in leckeren Schockwellen nach Norden, dann beginnt – ganz grob – mit der deutschen Sprache die Welt der deftigen Soßen und nahrhaften Eintöpfe. Und wenn man sich mit aller Leidenschaft durch die Töpfe geschlemmt hat, ist am Ende eben doch ein Muster erkennbar: Hauptsache lecker.

Es sind übrigens gefüllte Nudeltaschen, die uns von Ost nach West begleiten: Als Jiaozi oder Wan Tan werden sie in China zu ersten Reise-Kameraden, wechseln dann unterwegs immer wieder einmal sowohl die Größe als auch ihre Füllung und heißen nun ganz lange Pelmeni, Manti oder Pierogi. Selbst die koreanischen Mandu oder die italienischen Ravioli und Tortellini scheinen Verwandte dieses Alte-Welt-Fast-Food-Klassikers zu sein – in Österreich bringt die kulinarische Völkerverständigung »Schlutzkrapfen« auf den Teller und bei Porsche in Stuttgart werden »Maultaschen« serviert.

We experienced a world we wouldn't have dared to imagine in our wildest dreams. Sometimes, our minds linger on the rough, rugged, and wild jungle roads of Laos. Then we remember crossing the border into China, the can-do atmosphere of that country, its explosive growth, the vibrant and colorful cities that reach high into the sky. Glittering high-rise skylines in the dazzling nightlights of megacities we'd never heard of before.

Then, all at once, you find yourself in Europe. A bridge in Türkiye takes you out of Asia with astounding speed. We need to allow ourselves a little time for the final leg, deviating from the direct route and sneaking in through the back door to our destination. We travel through the rugged terrain of northern Greece and the wild regions of Bulgaria and Romania, tracing a wide arc through Hungary back to the Balkan countries. Europeans say that it is here that you truly feel how East becomes West, a completely different facet of humanity. But the truth is connecting these worlds is really quite easy. You just need to drive.

By the way, for all those who, for whatever reason, can only ever dream of making the long journey between Asia and Europe, we can recommend an alternative route. It begins in the wonderfully aromatic street food stalls of Southeast Asia and then makes its first stop in the incredibly diverse cuisines of China's regions. China is a culinary continent. The rest of the journey will suit meat-lovers well – vegetarians or even vegans will have a more challenging time in the vastness of Kazakhstan, and this culinary atmosphere, fragrant with lamb fat and heavy with the aroma of grilled food, penetrates far into the West. This culture of grilled meats extends as far as the Balkans and Greece, where vegetables from both the Old and New Worlds begin to make their mark. A culinary superpower sends delicious shockwaves northward from Italy and then, very roughly speaking, with the advent of the Germanic languages, the world of hearty sauces and nourishing stews begins. After you've indulged your hearty appetite, a pattern emerges: the main thing is that the food should taste good.

Incidentally, the culinary common denominator from East to West is filled pasta parcels. They become our first travel companions in China in the form of jiaozi or wonton dumplings, then changing size and filling along the way, eventually being called Pelmeni, Manti, or Pierogi. Even Korean mandu or Italian ravioli and tortellini seem to be relatives of this Old-World fast-food classic, while in Austria, good international culinary relations bring »Schlutzkrapfen« pasta pockets to the table, and »Maultaschen,« a close relative, is served at Porsche in Stuttgart.

Team

Tenn

Stefan

Aob

Oou

Joke

Humphrey

Suthi

Ake

A GT

Jack

Lit

Gap

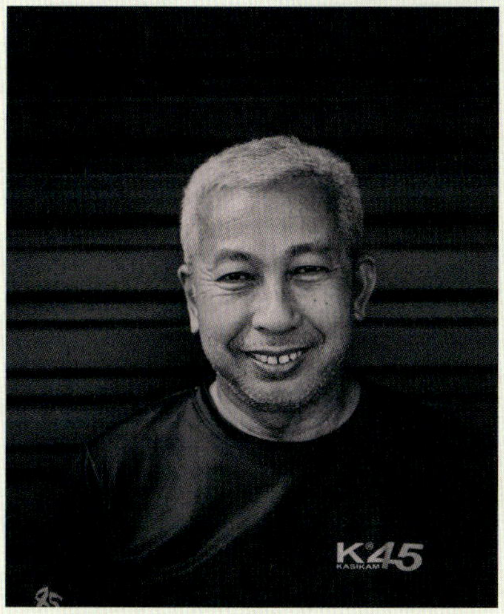

Mog

Chayanin

Art

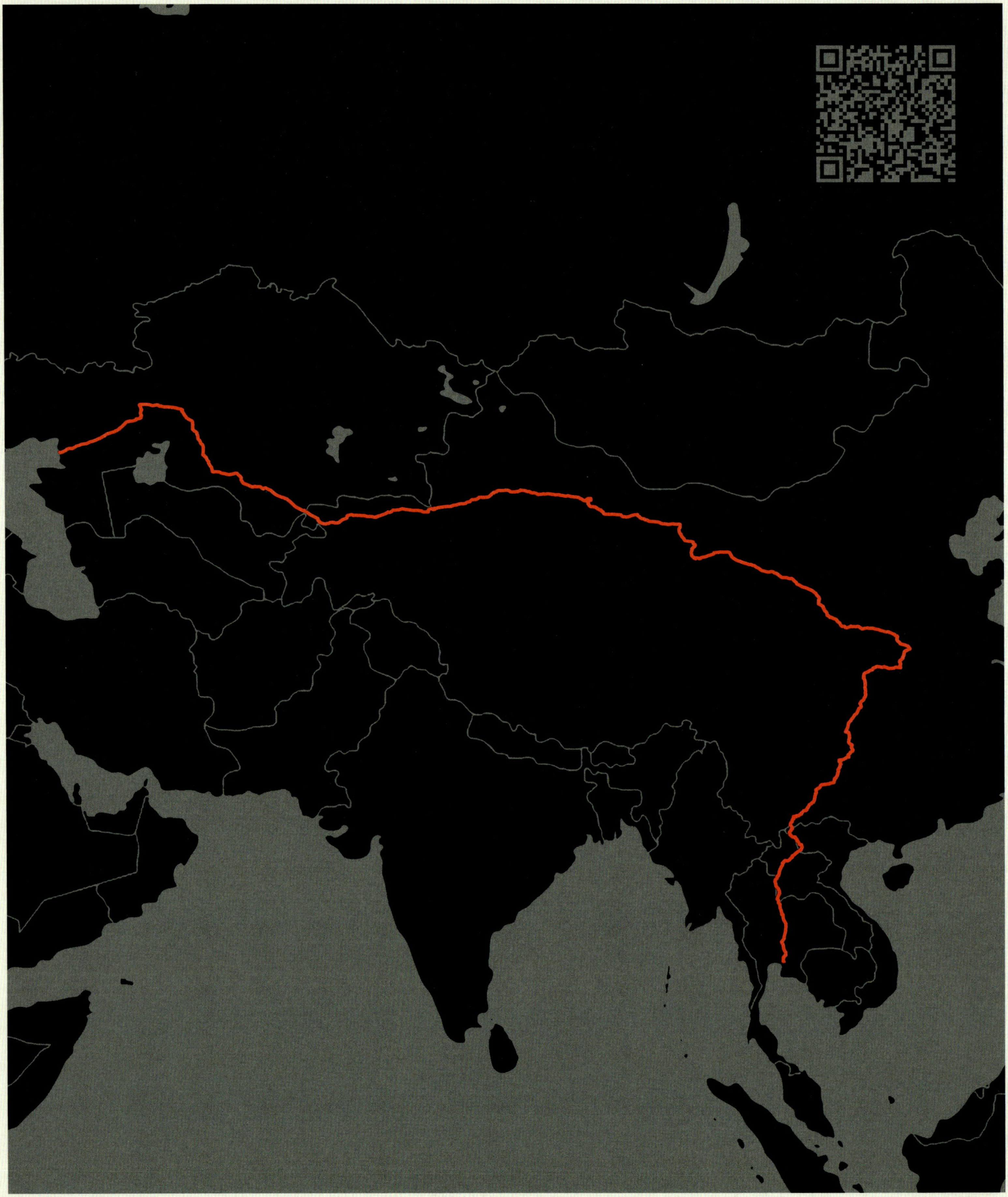

Etappe eins
Stage One

THAILAND / LAOS / CHINA / KASACHSTAN
THAILAND / LAOS / CHINA / KAZAKHSTAN

Ping. Ping. Ping. – Hohe, klare Frequenzen, flirrend über einer irdenen Grundnote ... Ein klein wenig hängt uns das sanfte Tönen der Klangschalen noch in den Ohren, genau wie der Duft von Räucherstäbchen in den Kleidern – Sandelholz, Myrrhe, Patchouli, Weihrauch. Selbstverständlich haben wir uns vor der bevorstehenden Reise auf diese Weise in der spirituellen Welt abgemeldet, und an ihren Schreinen Beistand für die kommenden Wochen und Hunderte, Tausende Kilometer erbeten. Seltsam. Selbst im säkularen Technik-Zeitalter ist einem irgendwie klar, dass die Welt aus mehr als dem Sichtbaren gestrickt ist. Sondern auch aus dem Unabwägbaren. Dem Schicksal. Dem Unausweichlichen. Und Nein, wir rechnen vor unserer großen Reise nicht damit, Glück zu haben – aber wir rechnen damit, glücklich zu sein. Und wenn man sich im euphorischen Schnattern umhört, das auf den ersten Kilometern der Fahrt blechern aus den Funkgeräten dringt, ist einem völlig klar: Glücklich sind wir schon jetzt. Und wie!

Der bittersüße Schmerz des Abschieds von Familie und Freunden steckt allen noch in den Gliedern, wir sind wohlwollend entlassen und von positiver Energie begleitet. Das Fernweh zieht uns mit ungeheurer Macht – und für die Zurückgebliebenen sind wir mit guten Wünschen, Hoffnungen und Träumen beladene Papierballons, die leuchtend in einen dunklen Nachthimmel aufsteigen. Auch ihre Abenteuerlust folgt uns. Treibt zusätzlich an. Als ob das überhaupt noch möglich wäre, denn unsere Gravitationsenergie ist immens: In einem zwingenden, schwerelosen, unerhört weit hinausführenden Parabelflug zieht es uns davon. Drei, zwei, eins – los!

Dabei war zu Beginn alles nur eine Idee. Die immer mehr Realität wurde und in Jahre der abwägenden, sorgfältigen Planung überging. Dann waren es nur noch Monate, schließlich Wochen, endlich Tage und jetzt sind wir unterwegs. Einfach so. Seltsam, wie banal Reisen beginnen, selbst wenn sie einen bis ans andere Ende zweier Kontinente führen: Zündung an, ersten Gang einlegen oder Wählhebel auf »Drive« schieben, dann auf die Straße hinaus. Genau wie sonst immer, wenn man zur Arbeit fährt, ins Restaurant, zur Familie und zu Freunden oder für ein paar Stunden ins Fahrfreude-Wochenende.

In Gedanken sind wir alles bereits vor dem Start abgefahren – wenn schon Langstrecke, dann richtig, haben wir uns gesagt.

Ding. Ding. Ding. – Shimmering frequencies float high above a deep, earthy note... The soft sound of the singing bowls still rings a little in our ears, just as the scent of joss sticks suffuses our clothing – sandalwood, myrrh, patchouli, incense. Ahead of the upcoming journey, it is only natural that we should sign off in this way in the spiritual world, and petition at the local shrines for assistance over the coming weeks and hundreds, thousands of kilometers. Strangely, even in this secular age where technology is king, it is somehow clear that the world comprises more than the visible and tangible. There is also the imponderable. Fate. The inevitable. And no, we don't expect to have more than our fair share of luck before our big trip – but we do hope to be happy. And if you ask around in the tinny euphoric chatter heard from our radios on the first few kilometers of the journey, one thing is completely clear: we are already genuinely happy!

We are all still feeling the bittersweet pain of saying goodbye to family and friends, who wave us off benevolently with plenty of positive energy. Wanderlust exerts its tremendous power – and for those left behind, we are like paper sky lanterns, laden with good wishes, hopes and dreams, shining in a dark night sky. Their thirst for adventure also accompanies us as an extra driving force if that were still possible. After all, the gravitational pull we ourselves experience is already immense, as we are whisked away in a compelling, weightless, outrageously farflung parabolic arc. Three, two, one – go!

In the beginning, it was all just a hare-brained idea. Gradually it gained more and more reality, turning into years of deliberate, careful planning. Then it was only a matter of months, then weeks, then finally days and now here we are on the road. Just like that. It's strange how banal journeys feel at the start, even if they are to take you to the other end of two continents: ignition on, engage first gear or push the selector lever to »Drive«, then out onto the road. Same as always, when you set off for work or a restaurant, drive to see family and friends or spend a few hours of driving pleasure at the weekend.

In our minds, we had already driven every inch of the journey before the start. We promised ourselves that if we were going to go long-haul, then we would do it properly. First Thailand, then Laos, and on to the border with China. From this point on,

Zuerst also Thailand, dann Laos, und weiter bis zur Grenze nach China. – Ab hier könnte das Vorankommen langwierig werden, denn nun bauen sich im Westen die Berge Tibets auf, die eine Routenwahl entlang der Luftlinie regelrecht unmöglich machen. Es müsste also mit Kurs Nord-Nordost an Chengdu vorbeigehen und erst, wenn man bei Xi'an landet, wäre an einen Kurswechsel in Richtung Westen überhaupt zu denken. Hier stehen dann schon weit über 3.000 Kilometer auf dem Tacho und die Fahrt hätte gerade erst begonnen. Es würde nun in die trockenen Hochebenen der chinesischen Provinz Gansu hinausgehen, die legendären Millionenstädte Chinas verteilen sich ab hier immer spärlicher und wenn man Jiayuguan erreicht hat, ist man ans westlichste Ende der alten chinesischen Reiche gelangt. Aber noch lange nicht ans Ende des heutigen China. Und man hätte sich dort gerade einmal den wohl herausforderndsten Teil der Reise nach Westen erschlossen – den Einzugsbereich der zentralasiatischen Grassteppen, das uigurische autonome Gebiet Xinjiang zwischen den Wüsten und dem Altai im Norden sowie dem Tian-Shen-Gebirge im Süden.

Spätestens an diesem Punkt wurde unser Finger auf der Landkarte tatsächlich unsicher und zaghaft: Eine solche Reise in der Theorie zu unternehmen, ist eben weitaus einfacher als jeden Kilometer wirklich zu fahren. Erstaunlicherweise hat aber genau diese Überlegung erst den eigentlichen Impuls zur tatsächlichen Reise gegeben: Man würde wohl einfach hinfahren müssen, um zu sehen, wie es sich anfühlt, wenn die Welt zu purer Distanz wird. Wie es ist, so entsetzlich weit weg von zu Hause zu sein, dass ein Umkehren sich nicht mehr lohnt. Wir denken an die antiken Reisenden auf der Seidenstraße, die auf dieser Route unterwegs waren. Die Karawanen und die Gesandten, die Sinn- und Weisheit-Suchenden, die Unruhegeister und Abenteurer. Wir wollen zu ihnen gehören. Und deshalb fahren wir. Genau jetzt und heute. Hinaus aus Bangkok, auf der AH1-Schnellstraße nach Norden, ein Pulk von Porsche-Freunden folgt uns. So müssen sich die Weltumsegler vergangener Jahrhunderte gefühlt haben, wenn sie den Hafen verließen: Fischerboote und Ausflugsdampfer im Schlepptau, winkend und Horn-tönend, bis auf die hohe See hinaus. – Selbstverständlich erzählen wir den Helden unseres begeisterten Abschieds-Konvois nicht von diesem Vergleich, die kosten schließlich gerade ihren eigenen, kleinen Weltenbummler-Moment aus. Spielen viellcicht sogar mit dem verwegenen Gedanken, sich doch noch anzuhängen, einfach so.

progress might be slow, because now the mountains of Tibet rise up to the west, which makes it virtually impossible to choose a route as the crow flies. So we would have to head north-northeast past Chengdu and only when we landed at Xi'an could a change of course to the west be contemplated. By this point we would have already clocked well over 3,000 kilometers on the odometer and our journey would have just begun. We would now venture into the dry plateaus of the Chinese province of Gansu. The legendary millionaire cities of China are spread out more and more sparsely from here on and when you reach Jiayuguan, you have reached the westernmost end of the old Chinese dominions. But we would still be far from the end of modern-day China, on the threshold of what is probably the most challenging part of the journey to the west – the Central Asian grassy steppes, the autonomous Uyghur region of Xinjiang between the deserts and the Altai in the north, and the Tian Shen Mountains in the south.

By this point at the latest, our confidence begins to dwindle as we study the map: undertaking such a journey is much easier in theory than actually driving every last kilometer. Surprisingly, however, it was exactly this thought that spurred us on to actually undertake the journey: You would just have to go there to find out how it feels when the world becomes pure distance and what it's like to be so far away from home that turning back is no longer a reasonable option. We think of the ancient travelers on the Silk Road who followed this route – the caravans and envoys, the seekers of meaning and wisdom, the restless spirits and adventurers. We want to be one of them. That's why we're driving. Right here and now.

We leave Bangkok on the AH1 expressway to the north with a band of Porsche fans in tow. This is how the seafaring navigators of past centuries must have felt when they left harbor: fishing boats and excursion steamers following in their wake, waving and sounding horns, all the way out to sea. Of course, we don't tell the heroes of our enthusiastic farewell convoy about this comparison, after all, they are just savoring their own little globetrotting moment. Who knows? Maybe some of them are even toying with the bold idea of joining the gang, just like that. All it takes is a short call to the family and the office: »I'm going to be away for a few days...« However, these pipedreams seem to come to an end as we reach the intersection with the AH2, and the

Hinaus in den grünen Norden Thailands, vorbei an Phitsanulok, in Richtung Uttaradit – und dort wechselt die Straße zum ersten Mal ihren Rhythmus. Das gleichmäßige Singen der Reifen auf dem Schnellstraßen-Asphalt wird nun ebenso unterbrochen, wie das gelangweilte Knurren der Motoren im letzten Gang. Kurven schneiden sich in den Dschungel, Stromleitungen begleiten die Straße – Drähte im stetig wippenden Auf-und-Ab-Groove zwischen den Masten. Leuchtend bunte Reklametafeln segeln uns am Straßenrand entgegen, huschen dann vorbei. Kleinstädte und Siedlungen tauchen auf, Hunderte Straßenschilder lotsen uns in alle Richtungen, aber wir bleiben auf Kurs.

Kurzer Anruf bei der Familie und im Büro: »Ich bin mal ein paar Tage weg ...« – Spätestens am Dreieck mit der AH-2 scheinen diese Gedankenspiele aber zu enden, die Abschieds-Komitees drehen winkend ab. Jetzt sind wir allein unterwegs. Fahren nach Norden. Immer weiter. Vorbei an der rund 50 Kilometer weiter nördlich gelegenen Großstadt Ayutthaya und bei Ang Thong haben wir dann die Ebenen am Chao-Phraya-Fluss erreicht, folgen ihm eine kleine Weile.

Aber nur solange, bis er in Richtung Westen abdriftet, während unsere Kompassnadel stoisch nach Norden zeigt. Hinaus in den grünen Norden Thailands, vorbei an Phitsanulok, in Richtung Uttaradit – und dort wechselt die Straße zum ersten Mal ihren Rhythmus. Das gleichmäßige Singen der Reifen auf dem Schnellstraßen-Asphalt wird nun ebenso unterbrochen, wie das gelangweilte Knurren der Motoren im letzten Gang. Kurven schneiden sich in den Dschungel, Stromleitungen begleiten die Straße – Drähte im stetig wippenden Auf-und-Ab-Groove zwischen den Masten. Leuchtend bunte Reklametafeln segeln uns am Straßenrand entgegen, huschen dann vorbei. Kleinstädte und Siedlungen tauchen auf, Hunderte Straßenschilder lotsen uns in alle Richtungen, aber wir bleiben auf Kurs. Die Giebel der Tempel Thailands zerfurchen den Himmel, der auch heute wieder die typische Farbe trägt: Blasses Blau, dunstige Wolken, Hitze und Luftfeuchtigkeit malen ein graublaues Aquarell in unzähligen Schattierungen.

Immer wieder wird unsere Karawane von Rollerfahrern begleitet, die in den Pulk eindringen, gespannt die bunten Sticker studieren, mit denen wir unsere Maschinen reisefertig gemacht haben. Irgendwann verstehen diese summenden Satelliten kopf-

farewell parties turn away with a wave. Now we are on our own. We head north, driving on and on. Passing the city of Ayutthaya, about 50 kilometers further north, we reach Ang Thong and the plains of the Chao Phraya River, which we follow for a short while, before it starts to drift westwards, while our compass needle steadfastly points north. We venture out into the green north of Thailand, past Phitsanulok, heading for Uttaradit, where the road changes its rhythm for the first time. The steady singing of the tires on the highway asphalt is interrupted, as is the bored growl of the engines in top gear. Curves slice into the jungle, with power lines flanking the road in a constant up-and-down groove between the pylons. Brightly colored billboards flash past us on the side of the road. Small towns and settlements appear, hundreds of street signs guide us in all directions, but we stay resolutely on course. The gables of Thailand's temples furrow the typically pale blue sky dotted with hazy clouds, where the heat and humidity paint a watercolor in countless shades of gray-blue.

Our convoy is frequently accompanied by scooter riders, inserting themselves between our cars and eagerly studying the colorful stickers we have added to our vehicles in preparation for travel. At some point, shaking their heads, honking their horns and gesticulating, these buzzing two-wheelers finally grasp where we come from and that we are bound on a journey far beyond their world, even though we're not sure that they can really estimate how far this really is. We kind of feel the same way too: theoretically, our route is set in marble, however in practice you just complete one kilometer after the other, letting the time pass. Each moment follows another.

Of course, because the journey is still so young, each of our traveling companions deals with these first few kilometers in their own particular way: there are the planners and strategists who like to think ahead and guess – or want to know – what lies in store for us. When is a good time for a pit stop for both driver and vehicle? What is the best, fastest and safest route? Will there be problems at the borders? Will the people be friendly, or will we face situations that overwhelm us and make the journey a failure? – It's good that there are people in the team who think ahead, as they bring dynamism to the flow, dragging us forward and weighing up the route before we ride it. But it's also good

schüttelnd, hupend und gestikulierend, woher wir kommen und dass es uns weit hinaus in die Welt tragen wird – ob sie wirklich abschätzen können, wie weit das in Wahrheit ist, wissen wir nicht. Und wir ahnen, dass es uns ähnlich geht: Theoretisch steht die Strecke, aber in der Praxis fährst du einfach einen Kilometer nach dem anderen, lässt die Zeit vergehen. Moment folgt auf Moment.

Und natürlich geht jeder der Reisegenossen auf seine Weise mit diesen ersten Kilometern um, auf denen die Reise noch so jung ist: Da sind die Planer und Strategen, die gern vorausschauen und erahnen – oder besser: wissen wollen –, was auf uns zukommt. Wann bietet sich ein Boxenstopp für Mann und Maschine an? Was ist die beste, schnellste, sicherste Route? Wird es Probleme an den Grenzen geben? Werden wir auf wohlwollende Menschen treffen, oder in Situationen geraten, die uns überfordern und die Reise scheitern lassen? – Gut, dass es diese Vordenker im Team gibt, sie bringen Dynamik in den Flow. Ziehen vorwärts. Denken die Route, bevor wir sie fahren. Aber gut auch, dass es die Genießer und Moment-Menschen im Team gibt. Die mit großen, neugierigen Augen in eine Freude hineinfahren, die wiederum alle anderen ansteckt. Eine kleine Stimme im Hinterkopf sagt dir, dass da draußen Momente auf uns warten, in denen wir die Unbekümmertheit und Leidenschaft dieser Reisegenossen dringender benötigen als alles andere … Und in diesem Bewusstsein fahren wir einen ganzen Tag nach Norden. Wir werden zum Wolfsrudel, das gierig und fokussiert dahinschnürt. Wir rechnen längst nicht mehr in einem, fünf oder zehn zurückgelegten Kilometern, sondern in 100 Kilometern. Und vielleicht stimmt nicht einmal das, vielleicht sind wir längst in einem Zustand der Meditation unterwegs: alle Antennen auf Empfang, die Farben und Gerüche und Stimmungen des Lands einsaugend.

Bevor wir nach Laos hineinfahren, legen wir noch eine nicht ganz freiwillige Belastungs-Generalprobe ein: Nins luftgekühlter 911 Safari verröchelt mit schleifender Kupplung am Straßenrand. Wenn wir es jetzt schaffen, den alten Racer wieder ans Laufen zu bekommen, dürfte Ähnliches auch anderswo funktionieren. Genau jetzt haben wir also bereits die Sollbruchstelle erreicht. Den Moment, in dem wir feststellen werden, ob unsere Planungen aufgehen … Die Reparatur wird zur Teamleistung: Wäh-

We venture out into the green north of Thailand, past Phitsanulok, heading for Uttaradit, where the road changes its rhythm for the first time. The steady singing of the tires on the highway asphalt is interrupted, as is the bored growl of the engines in top gear. Curves slice into the jungle, with power lines flanking the road in a constant up-and-down groove between the pylons. Brightly colored billboards flash past us on the side of the road. Small towns and settlements appear, hundreds of street signs guide us in all directions, but we stay resolutely on course.

that the team has its share of people who just enjoy life and what it brings at any given moment. These people drive with big, curious eyes, and their enthusiasm infects everyone else. A little voice in the back of your mind tells you that there are moments out there waiting for us when we will need the carefree passion of these travel companions more than anything else...

With this in mind, we travel north for a whole day. We are like a greedy and focused wolf pack. We no longer calculate the distance traveled in one, five or ten kilometers covered, but in hundreds of kilometers. Perhaps not even that is true. Maybe we have long been in a state of meditation: all antennae are tuned to receive, absorbing the colors, smells, and moods of the land.

Before we cross into Laos, we complete a not entirely voluntary stress test: Nin's air-cooled 911 Safari is lying on the side of the road suffering from clutch problems. If we can manage to get the venerable racer up and running again, then we should be able to achieve something similar elsewhere should the need arise. So right now, we have already reached the predetermined breaking point, the moment when we find out whether our plans will work... The repair becomes a team effort: while the confident-looking support crew jacks up the 911, lifts out the engine and transmission and changes the clutch in a kind of live open-heart surgery in front of an audience, the rest of the team fans out in search of street food. They return with spicy soups, grilled poultry, fruit, and sweets, along with a healthy dose of good

rend die souverän agierende Support-Crew den 911 aufbockt, Motor samt Getriebe ausbaut und in einer Art Live-OP am offenen Herzen vor Zuschauern die Kupplung wechselt, schwärmt der Rest der Mannschaft auf der Suche nach Streetfood aus. Schleppt scharfe Suppen an, gegrillte Geflügelteile, Obst und Süßigkeiten, samt guter Laune, Zuversicht und Weiter-so-Parolen. Ein klein wenig ist das, wie wenn Ameisen im Unterholz den Kadaver einer Maus finden, ihn langsam zerlegen und in den Bau schleppen – nur andersherum: Der ausgeweidete 911 wird zum Mittelpunkt einer emsig werkenden Mannschaft inklusive Versorgungs-Infrastruktur. So lange, bis die knallgelbe Maschine wieder auf ihren Rädern steht und nach einem Dreh am Zündschlüssel mit unternehmungslustig kläffendem Motor sowie perfekt zupackender Kupplung davonprescht. Werkzeug einsammeln, Ersatzteil-Boxen nachsortieren, Abfahrt zur Grenze nach Laos.

Der mächtige Mekong ist die natürliche Barriere zwischen den Staaten. Bevor man ihn auf der »Friendship-Bridge« überqueren darf, muss man durchs Nadelöhr der Grenzkontrollen. Und die fallen für ein buntes, lautes Rudel von Porsche-Sportwagen nicht unbedingt strenger aus als sonst, aber interessierter, erstaunter, verwunderter. Wenn man auf die Frage nach dem »Wohin« mit »zuerst nach China, eigentlich aber nach Deutschland« antwortet, überlegt so ein Grenzbeamter definitiv lang und sorgfältig, ob er gerade auf den Arm genommen werden soll. Fünf Autos und zwei Support-Crews später glaubt er ganz sicher an die Ernsthaftigkeit unseres Vorhabens, und nachdem er beim ersten Mal noch einen prüfenden, unsicheren Blick auf die Weltkarte im Büro geworfen hat, um sicherheitshalber nachzusehen, wo dieses »Deutschland« eigentlich ist, weiß er es nach dem letzten Auto ganz bestimmt: »Ach, Deutschland, klar. Rüber nach China, dann Kasachstan, und dann ist es nicht mehr weit, oder?« – Ja. So ungefähr …

Jetzt aber erst einmal: Laos. Und das bedeutet: einen schlagartigen Downgrade in Sachen Straßenqualität. Während uns Thailand mit teilweise delikatem Streetsurfing-Asphalt versorgt hat, gut ausgebauten Schnellstraßen und mehrspurigem Fast-Forward-Material, geben die rund 230 Kilometer von der thailändischen Grenze bis nach Boten, an der Grenze zu China, so ziemlich alles, um uns auf ein Leben diesseits des vierten Gangs

humor, confidence and encouraging words. It's a little bit like when ants find the carcass of a dead mouse in the undergrowth, slowly dissect it and drag it back to their nest – except the other way around: the gutted 911 becomes the focus of attention for a busily working team and their supply infrastructure. Eventually the bright yellow machine is put back on its wheels and, after a turn of the ignition key, shoots off with an excitable, roaring engine and a perfectly gripping clutch. Tools are collected, boxes of spare parts get sorted, and we set off for the border with Laos. The mighty Mekong River is the natural barrier between the countries. Before you are allowed to cross it on the »Friendship Bridge«, you must pass under the gimlet eye of the border control officers. The customs officials aren't necessarily stricter than usual when faced with a colorful, loud pack of Porsche sports cars, but they are certainly more interested, more surprised, more curious. If when asked where you're going you answer »first to China, but actually our final destination is Germany«, any reasonable border guard will think long and hard about whether you're pulling his leg. Five cars and two support crews later, he is certainly convinced of the seriousness of our project. After taking a scrutinizing, uncertain look at the world map in his office for the first time to be on the safe side, he knows the answer for sure after the last car: »Oh, Germany, yes of course. Over to China, then Kazakhstan, and then it's not far, is it?« – Yes. Something like that...

For now, let's focus on Laos. And that means: a sudden drop in terms of road quality. While Thailand has provided us with some delicate street-surfing asphalt, well-developed main roads and fast-paced multi-lane highways, the approximately 230 kilometers from the Thai border to Boten, on the border with China, throws pretty much everything at us to get us in the mood for a life this side of fourth gear. At best, the road meanders incessantly through the mountains of the Laotian north as a narrow ribbon of asphalt, but most of the time it ekes out a meager pothole-ridden existence. However, the word »pothole« shouldn't be taken too literally – after all, the term comes from a world in which a continuous asphalt surface is the norm and potholes are the exception. Here, on state road number 3 from Ban Houayxay to Boten, however, the situation is exactly the opposite: colonies of potholes merge into one another to become the actual default state, interrupted by compacted clay or occasional

While Thailand has provided us with some delicate street-surfing asphalt, well-developed main roads and fast-paced multi-lane highways, the approximately 230 kilometers from the Thai border to Boten, on the border with China, throws pretty much everything at us to get us in the mood for a life this side of fourth gear. At best, the road meanders incessantly through the mountains of the Laotian north as a narrow ribbon of asphalt, but most of the time it ekes out a meager pothole-ridden existence.

einzustimmen. Im besten Fall schlängeln sie sich als schmales Asphaltband unaufhörlich durch die Berge des laotischen Nordens, meistens fristet die Straße aber ein vernachlässigtes, von Schlaglöchern übersätes Dasein. Wobei man »Schlagloch« keineswegs wörtlich nehmen sollte, der Begriff stammt schließlich aus einer Welt, in der eine durchgehende Asphaltierung Normalzustand ist, und Schlaglöcher singuläre Ereignisse sind. Auf der Staatsstraße Nummer 3 von Ban Houayxay nach Boten verhält es sich jedoch immer wieder genau umgekehrt: Ineinander übergehende Schlagloch-Kolonien werden zum eigentlichen Aggregatzustand, unterbrochen von glatt gefahrenem Lehm oder gelegentlichen Asphaltpassagen. Wir kurven also. Die ganze Zeit. Über 200 Kilometer weit. Weil die Straße eh nicht gerade verläuft, sondern sich wild durch atemberaubend schöne Gebirgslandschaften schlängelt, oder weil man besser einen großen Bogen um die Löcher macht, deren tückische Tiefe selten zu sehen, aber hart zu spüren ist.

Gut, dass wir mit dem Cayenne Coupé, 911 Dakar und Ous hochgelegtem Cayman gleich drei ausgemachte Schlechtwege-Experten fahren. Der leuchtend gelbe Safari-911, in dem Chayanin unterwegs ist, macht uns ohnehin keine Sorgen, das Kupplungs-Update läuft wie geschmiert. Nur Humphreys Flüche aus dem vipergrünen 992 GTS knarzen laut im Funkgerät, die tief kauernde Anpressdruck-Konfiguration des schnellen Sportlers lässt nun die Anschlagdämpfer rumsen und die Wirbelsäule knacken. Für Laos brauchen wir ganze zwei Tage, statt der erwarteten 5–6 Stunden. Und wir lernen dabei: Auf Fernreisen nimmt man die Stoppuhr besser nicht so ernst. Sondern lässt sich auf das ein, was einem serviert wird. Heiß, lecker und scharf, eben asphaltiert oder löchrig, seidenweich oder eben ölig und schleifend. Alles gehört zusammen. Ohne Schatten ist das Licht manchmal überhaupt nicht zu sehen.

Am Grenzübergang nach China sind wir Gereifte. Fühlen uns wie alte Weitreise-Hasen, für alles gewappnet. Allerdings rechnen wir für den Eintritt unserer Karawane in die Volksrepublik trotz aller Reisevorbereitungen mit Überraschungen – und sind mehr als überrascht über die schnelle, effiziente Abfertigung: Um in China fahren zu dürfen muss eine kurze Einweisung sein, sozusagen eine kleine Fahrprüfung im Schnellverfahren. Sobald die absolviert ist, werden uns »provisional driving permits« über-

passages of asphalt. So we need to curve all the time for over 200 kilometers. The road doesn't run straight anyway, but meanders wildly through breathtakingly beautiful mountain landscapes, or because it's better to give the holes a wide berth, their treacherous depths rarely seen, but felt to the core of your bones. Happily, we are driving three tried-and-tested experts in poor surfaces in the form of the Cayenne Coupé, 911 Dakar and Ou's high-clearance Cayman. The bright yellow Safari 911 driven by Chayanin is no longer a worry anyway, as the clutch replacement runs like clockwork. Only Humphrey's curses from behind the wheel of the viper green 992 GTS sound loudly from the radio. The low-lying downforce configuration of this superfast sports car now causes the stop buffers to groan and the vehicle's spine to crunch. We need a whole two days for Laos, instead of the expected 5 to 6 hours. We learn something along the way: it's better not to watch the clock too seriously on long-distance trips. Instead, you learn to accept what you get, whether hot, tasty and spicy, evenly paved or full of holes, soft as silk, or even gritty and greasy. It is all one. Sometimes you need shadows to see the light.

By the time we reach the border crossing to China, we have attained a new level of maturity. We feel like old hands at long-distance travel – prepared for anything. However, despite all the travel preparations, we expect some surprises when our caravan tries to enter the People's Republic. In fact, we are astonished by the speed and efficiency we experience: in order to be allowed to drive in China, we need to complete a short briefing, kind of like a fast-track mini driving. As soon as this is done, we get »provisional driving permits«, allowing us to set off on the upcoming 5,500-kilometer journey. We are also given special short-term

> An den Abenden finden wir uns an einem enormen Tisch wieder, um eine zweite Reise anzutreten: Essen. Bunt und aufregend, fettig und salzig, süß und sauer, mild und scharf, schleimig und texturreich. Wir erkennen die Grundzutaten, haben sie aber noch nie in einer ähnlichen Form erlebt. Am Anfang sitzt du vor diesem Sortiment von fermentiertem Gemüse, eingelegten Pilzen, gekochten Eiern, gefüllten Teigtaschen oder -bällchen, Gegrilltem und Gedünstetem, und du weißt einfach nicht, was nun von dir erwartet wird. – Essen?

reicht, mit denen wir uns auf die bevorstehenden 5.500 Kilometer machen können. Spezielle Kurzzeitkennzeichen gibt es ebenfalls – und dann stehen wir plötzlich in China. Willkommen und Gute Fahrt!

Das Land hebt uns völlig aus den Angeln, stürzt uns von einer Überraschung in die nächste. Selbst hier unten im Süden mit seinen üppigen Bergwäldern, ist das Tempo enorm. Wie mit Lineal und Zirkel gezogene Schnellstraßen schieben selbstbewusst durchs Land, Brücken und Tunnel hauen sich beinahe ungerührt von der Topografie vorwärts. Wir machen regelrecht rauschhaft Kilometer, werfen uns aber auch kopfüber ins China-Erlebnis. Während der Tage fahren wir mächtige Etappen, bei mildem Sonnenschein oder tosendem Regen, die ganze Zeit in einem Zustand latenter Überforderung: Das Maß an Neuem, nicht Interpretierbarem ist immens. China überschüttet uns mit der Diskrepanz von ländlicher Weite und beinahe überdrehter Moderne, selbstverständliche Traditions-Elemente treffen auf ungerührte Digitalität. Einen Blick in den Rückspiegel scheint hier niemand zu werfen, es geht ausschließlich und mit hohem Tempo nach vorn. Atemberaubend und beängstigend ist das, aber auch ansteckend, animierend, begeisternd.

An den Abenden finden wir uns an einem enormen Tisch wieder, um eine zweite Reise anzutreten: Essen. Bunt und aufregend, fettig und salzig, süß und sauer, mild und scharf, schlei-

license plates – and then suddenly we find ourselves in China. Welcome and have a good trip!

We are completely bowled over by the country, plunging from one surprise to the next. Even down here in the south with its lush mountain forests, the pace of driving is enormous. Highways that seem to have been drawn with a ruler and compass push confidently through the countryside, bridges and tunnels smash forward with little concern for the topography. Intoxicated by the chance to cover so much ground, we throw ourselves headlong into the China experience. Whether in mild sunshine or pouring rain, we complete mighty stages during the day in a constant state of latent sensory overload: the number of new, mystifying sights we witness is immense. China overwhelms us with the disconnect between its rural expanses and almost excessive modernity. Long-established traditions meet an uncompromising digital world. Nobody seems to look in the rear-view mirror here and the only possible direction is forward at breakneck speed. It's all quite breathtaking and frightening, but also infectious, exhilarating, and inspiring.

In the evenings, we assemble around an enormous table to embark on a second odyssey: food. The dishes we sample are colorful and exciting, fatty and salty, sweet and sour, mild and spicy, slimy and rich in texture. We recognize the basic ingredients but have never experienced them quite like this before. To begin with, you are confronted with such an assortment of fermented vegetables, pickled mushrooms, boiled eggs, stuffed dumplings or dough balls, grilled and steamed meats that you just don't know what is expected of you. Can this really be food? The explosion of strange colors and scents is very different from anything you know. Then, armed with chopsticks, spoons and fingers, you launch yourself into a colorful, exotic garden, fighting bravely until you recognize something. This will then become your comfort zone in the thicket of tastes, something you can always return to if it all gets too colorful for you. At least you have worked out one flavor, so you can continue from here... You find your trips to the different pots and bowls getting longer and longer. Your palate has begun to change, just like your expectations. You have reached the point of no return and are now ready for the really hard stuff: pickled lotus roots, for example, which you would find musty and slimy anywhere else, but rich and tender

mig und texturreich. Wir erkennen die Grundzutaten, haben sie aber noch nie in einer ähnlichen Form erlebt. Am Anfang sitzt du vor diesem Sortiment von fermentiertem Gemüse, eingelegten Pilzen, gekochten Eiern, gefüllten Teigtaschen oder -bällchen, Gegrilltem und Gedünstetem, und du weißt einfach nicht, was nun von dir erwartet wird. – Essen? Diese Explosion von fremden Farben und Düften, alles ganz anders, als du es kennst. Und dann gehst du mit Stäbchen, Löffeln, Fingern bewaffnet hinein in einen bunten, exotischen Garten, schlägst dich tapfer, bis du irgendetwas wiedererkennst. Das wird dann deine Komfort-Zone im Dickicht der Geschmäcker, hierher kannst du immer wieder zurückkehren, wenn dir alles zu bunt wird. Zumindest den einen Geschmack hast du dir erarbeitet, von hier aus kann es weitergehen … Du ertappst dich dabei, wie deine Ausflüge in die Töpfe und Schalen immer länger und weiter gehen. Wie sich dein Gaumen verändert, genau wie deine Erwartung. Hier ist der Punkt ohne Wiederkehr erreicht, jetzt bist du bereit für die ganz harten Sachen: eingelegte Lotos-Wurzeln, zum Beispiel, die man anderswo muffig und schleimig finden würde, hier aber gehaltvoll und zart. Du lässt dir das feine Fett unter einer Entenhaut am Gaumen zergehen oder tunkst fein geschnittene Fleisch-Scheiben in brodelndes Fett, in dem lavascharfe Schoten blubbern. Und wenn du dir die dann kross frittierten Fleischstücke lässig auf die Zunge legst, stört dich überhaupt nicht, dass dein Mund brennt, am ganzen Leib Schweiß in Strömen ausbricht, die Welt im Feuer zergeht – du badest im Jetzt des Schlemmens mit all seinen Sensationen. Mit vollem Mund kauend schaust du zu, wie sich die Tischplatte an dir vorüberdreht, als Karussell des Fressbaren. China, du bist verrückt.

Manchmal erinnern wir uns zurück, an die leichten, filigranen Genüsse Thailands, aber das Gefühl verblasst. Wir sind unterwegs. Kauen und schlucken jeden Kilometer. Genauso wollten wir das: Tagsüber im Flow des Fahrens abtauchen und dann in die Nächte der Millionenstädte eintauchen, wenn LED-Comicstrips ganze Wolkenkratzer-Fassaden bunt erschauern lassen. Because we can. In China denkt man gerade groß, nimmt den ganz langen, tiefen Schluck aus der Was-kostet-die-Welt-Pulle und es ist mitreißend, dabei zuzuschauen. Die unbändige Lust an Aufwärts- und Vorwärts-Entwicklung wirkt aber immer wieder auch wie ein Rausch, bei dem man sich nicht ganz sicher ist, wie groß der Kater danach sein könnte. Seine besten Momen-

China is quite crazy. Sometimes we think back to the light, delicate pleasures of Thailand, but the feeling soon fades. We are on the move, chewing and swallowing every kilometer. This is exactly what we were after: the chance to dive into the flow of driving during the day and then immerse ourselves in the nightlife of the megacities, when LED comic strips make entire high-rise facades shimmer with color. Because we can.

here. You let the fine fat under the duck skin melt on your tongue or dip finely chopped meat slices in boiling oil where lava-hot peppers bubble away. When you casually drop the crispy fried morsels of meat onto your tongue, you don't mind at all that your mouth burns and sweat breaks out all over your body. The world is on fire and you bathe in the immediacy of a feast of sensations. Chewing with your mouth full, you watch as the tabletop spins past you, like a culinary carousel. China is quite crazy. Sometimes we think back to the light, delicate pleasures of Thailand, but the feeling soon fades. We are on the move, chewing and swallowing every kilometer. This is exactly what we were after: the chance to dive into the flow of driving during the day and then immerse ourselves in the nightlife of the megacities, when LED comic strips make entire high-rise facades shimmer with color. Because we can.

People in China are thinking big right now, drinking long and deep from the fountain of opportunity. Although this is thrilling to witness, the irrepressible desire for upward and forward development always seems like a rush, so you are not quite sure how big the hangover might be afterwards. This wonderful, confusing, driven country is probably at its best when it remembers the past, or pauses for a brief moment of introspection. This is when magical and picture-perfect emotions suddenly emerge. Bittersweet and fragile, profound and tender, occasionally profoundly sad. Each moment intensifies our love affair with the Chinese way of life, as we enjoy the country's dynamism, admiring its self-confidence and surprised by its diversity. Of course we get completely lost along the way. China is a whole universe unto itself, making it impossible to sort your impressions, or to understand this country. However, you can let yourself drift, like us: we skirt Pu'er on the way up to Kunming – a distance of almost

Bittersüß und zerbrechlich, tiefsinnig und zart, gelegentlich auch todtraurig. Wir werden mit jedem Moment mehr ein Fan des China-Kosmos, genießen die Dynamik, bewundern das Selbstbewusstsein, sind überrascht von Vielfalt. Und ja – wir gehen unterwegs komplett verloren. China ist ein ganzes Universum. Du kannst die Eindrücke unmöglich sortieren, dieses Land unmöglich verstehen. Aber man kann sich treiben lassen. Wie wir: vorbei an Pu'er, hinauf nach Kunming – allein das sind beinahe 700 Kilometer. Vorbei an Qujing, dann genau zwischen Chengdu und Chongquing hindurch, nach Xi'An.

te hat dieses wunderbare, verwirrende, getriebene Land daher auch dann, wenn es sich an die Vergangenheit erinnert oder einen kurzen Moment lang innehält. Dann kommen plötzlich zauberhafte und bildschöne Emotionen hervor. Bittersüß und zerbrechlich, tiefsinnig und zart, gelegentlich auch todtraurig. Wir werden mit jedem Moment mehr ein Fan des China-Kosmos, genießen die Dynamik, bewundern das Selbstbewusstsein, sind überrascht von Vielfalt. Und ja – wir gehen unterwegs komplett verloren. China ist ein ganzes Universum. Du kannst die Eindrücke unmöglich sortieren, dieses Land unmöglich verstehen. Aber man kann sich treiben lassen. Wie wir: vorbei an Pu'er, hinauf nach Kunming – allein das sind beinahe 700 Kilometer. Vorbei an Qujing, dann genau zwischen Chengdu und Chongquing hindurch, nach Xi'An.

2.100 Kilometer haben wir bis hierher in der Volksrepublik zurückgelegt und Xi'An wird zum Wendepunkt. Die erste Hauptstadt des »Vereinigten Kaiserreichs China« von weit vor der christlichen Zeitrechnung steht exakt zwischen dem alten und neuen China: Einerseits marschiert hier die Terrakotta-Armee des »Ersten erhabenen Gottkaisers von Qin« in gefühlt quicklebendigen Kolonnen direkt aus der Antike in die Postmoderne hinein, andererseits ist das brodelnde Xi'An von heute, mit seinen rund 13 Millionen Einwohnern, unser Abschied aus dem China, das wir bisher kennengelernt haben. Von hier aus geht es nach Westen. Hinaus in die riesigen Gebiete der Provinzen Gansu und Xinjiang, hinaus aus dem subtropischen Klima des Südens in die trockenen Steppen und Wüsten. Ab hier gehen uns die Multimillionenstädte aus. Ab hier regiert unermessliche Weite. Nun sind wir tatsächlich auf den Spuren der alten Seidenstraße unterwegs, nicht mehr als Weitreisende, sondern als Zeitreisende. Auf endlosen Straßen fährt jeder ganz allein

700 kilometers alone – then on past Qujing, passing between Chengdu and Chongqing, to reach Xi'An.

We have covered 2,100 kilometers so far in the People's Republic and Xi'An represents a turning point. The first capital of the »United Empire of China« long predates the Christian era and stands exactly on the fault line between old and new China. On the one hand, this is where the terracotta army of the »First August God-Emperor of the Qin Dynasty« marches in what feels like living serried ranks, stepping directly from antiquity into postmodernity. On the other hand, today's bustling Xi'An, with its approximately 13 million inhabitants, marks our farewell to the China we have come to know so far. From here we are heading west, out into the vast provinces of Gansu and Xinjiang, away from the subtropical climate of the south to a landscape of dry steppes and deserts. We run out of cities of millions to visit, as from here on, immeasurable expanses reign. We are actually following the old Silk Road, no longer as long-distance travelers, but as explorers in time. On these endless roads, each of us drives all alone, focused on their personal center point and finding their flow at that exact moment when they feel they are no longer moving.

Somewhere out there, we start to believe that our goal might actually be within reach, although it may be because the idea of arriving at our destination is becoming increasingly unreal and we are moving completely in the here and now. The countryside is enchantingly beautiful: first terraced fields cling to rugged hilly landscapes, then rocky river valleys lead into thorny, dust-dry steppes. Red crags line the road for days on end, turning our cars into spaceships sailing across the desert surface of Mars. We see temples carved into the mountains and statues of Buddha telling sublime tales, while next door, the Chinese government is building 20-gigawatt windfarms with 7,000 wind turbines in the desert, easily ten times the size of the largest wind farms in the US. This is a serious business: China is marching hurriedly onward toward tomorrow.

The culture also changes with the landscape: China's wild west is rough and barren and that rubs off on its inhabitants. This area is characterized by the equestrian and nomadic cultures of the Turkic and Mongolian peoples of Central Asia. Muslim elements slowly take over... Just as we have lost all sense of space

seinem persönlichen Totpunkt entgegen und findet genau dann, wenn er meint auf der Stelle zu stehen, seinen Flow.

Irgendwo dort draußen werden wir anfangen zu glauben, dass das Ziel tatsächlich in Reichweite geraten könnte – vielleicht auch, weil das Ankommen immer unwirklicher wird und wir uns ganz im Jetzt bewegen. Die Schönheit des Lands ist berückend: Zuerst wogen Terrassenfelder über schroffe Hügellandschaften, dann führen felsige Flusstäler hinaus in dornige und staubtrockene Steppen. Rote Felsen begleiten die Straße tagelang, die Autos werden zu Raumschiffen auf Wüstenplaneten, segeln über die Oberfläche des Mars. In die Berge gehauene Tempel und Buddha-Statuen erzählen erhabene Geschichten – und nebenan pflanzt die chinesische Regierung 20-Gigawatt-Windkraftwerke aus 7.000 Windrädern in die Wüste. Lässig mal die zehnfache Größe der größten US-Windfarm. Die meinen es ernst – da ist es wieder, das im Eiltempo nach Morgen marschierende China. Mit dem Land verändert sich auch die Kultur: Der Wilde Westen Chinas ist rau und karg und das färbt auf die Menschen ab. Die Reiter- und Nomaden-Kulturen der Turkvölker und mongolischen Völker Zentralasiens prägen diese Gegend, langsam übernehmen muslimische Elemente ... Gerade als wir jegliches Gefühl für Raum und Zeit verloren haben, tauchen schneebedeckte Berge im Süden auf, und dann haben wir plötzlich die Grenze zu Kasachstan erreicht. Sie ist einfach da, völlig unwirklich und schockierend, wirkt wie ein Star-Wars-Außenposten am Rand einer Galaxie. Die Straße steigt vorher über lange Zeit aus den trockenen Ebenen Xinjiangs immer weiter hinauf, landet auf einer Passhöhe, der europäische Reisende eine verblüffende Ähnlichkeit mit dem Tausende Kilometer und einen Kontinent weit entfernten San-Bernardino-Pass konstatieren. Von hier aus segeln wir wie in einen Wachtraum geraten bergab, zur Grenze.

Unsere Porsche-Karawane rollt geduldig durch die Prozeduren des Ein- und Auscheckens, stoische Geduld haben wir inzwischen über 6.000 Kilometer lang studiert. Und dann sind wir auf der anderen Seite, rollen unter den Schneegipfeln des Tian-Shan-Gebirges in Richtung Almaty. Zur Rechten erstrecken sich ungeheure Ebenen, zur Linken rollen mächtige Bergketten heran. Im ersten Moment fühlt sich das an, als hätten wir endgültig das Ende der Welt erreicht, aber dann wird uns klar, dass diese Ecke der Welt aus anderen Perspektiven betrachtet viel-

The culture also changes with the landscape: China's wild west is rough and barren and that rubs off on its inhabitants. This area is characterized by the equestrian and nomadic cultures of the Turkic and Mongolian peoples of Central Asia. Muslim elements slowly take over... Just as we have lost all sense of space and time, snow-capped mountains appear to the south, and we find we have suddenly reached the border with Kazakhstan. The settlement around the crossing looks completely unreal and quite astonishing, like a Star Wars outpost at the edge of a galaxy.

---

and time, snow-capped mountains appear to the south, and we find we have suddenly reached the border with Kazakhstan. The settlement around the crossing looks completely unreal and quite astonishing, like a Star Wars outpost at the edge of a galaxy. The road climbs painstakingly up and up from the dry plains of Xinjiang, arriving at a pass that, to European eyes at least, bears a striking resemblance to the San Bernardino Pass, thousands of kilometers and a whole continent away. From here we descend as if in a waking dream, to the border.

Our Porsche posse patiently completes the border formalities, displaying the stoic forbearance we have learned over 6,000 kilometers. We then find ourselves on the other side, cruising in the shadows of the snowy peaks of the Tian Shan Mountains towards Almaty. To the right lie immense plains, while to the left mighty mountain ranges set up camp. At first it feels like we've finally reached the end of the world, but then we realize that for someone else this corner of the globe may be more like the center of the universe: the area has been inhabited for almost 3,000 years, possibly starting with Persian-speaking nomads. Today's city has well over two million inhabitants. The name »Almaty« refers to the fact that the first early form of today's domestic apple was cultivated here, quite a mega claim to fame, all in all.

About 700 kilometers later, Shymkent marks the southernmost point of our journey through Kazakhstan. A little over 100 kilometers further south is Toshkent, the capital of Uzbekistan. This is the winter route for the Silk Road down to Samarkand, when ice-cold winter storms up in Kazakhstan smother the steppe under several meters of snow. However, we are taking the summer route, clawing our way across the planet kilometer-by-kilometer as a united team. Our engines have been running almost non-

leicht eher ihr Mittelpunkt ist: Seit fast 3.000 Jahren ist die Gegend besiedelt, den Anfang dürften persischsprachige Nomaden gemacht haben. Die Stadt hat heute weit über zwei Millionen Einwohner, der Name »Almaty« bezieht sich auf die Tatsache, dass hier die erste Frühform des heutigen Kulturapfels gezüchtet wurde. Wenn das keine globale Mega-Karriere ist, die hier ihren Ursprung hat.

Rund 700 Kilometer später wird Schymkent zum südlichsten Punkt der Fahrt durch Kasachstan, etwas über 100 Kilometer weiter im Süden liegt Taschkent, die Hauptstadt Usbekistans: Seidenstraße-Winterroute hinunter nach Samarkand, wenn oben in Kasachstan eiskalte Winterstürme die Steppe unter meterhohem Schnee ersticken. – Wir sind allerdings auf der Sommerroute unterwegs, ringen dem Planeten als eingeschworenes Team Kilometer um Kilometer ab. Die Motoren laufen nun schon seit Wochen beinahe Nonstop, ohne Probleme, selbst der Safari-Klassiker wettert hartnäckig vorwärts. Der Blick über die Fronthaube auf die im Staub voraus dahinziehenden Reise-Genossen ist ein Bild, das uns selbst nachts im Schlaf erscheint. Bei Baikonur werfen wir einen Blick auf die legendären Anlagen der sowjetischen Raumfahrt, während ringsum ein anderer Raum beinahe endlos wirkt. Die kasachische Steppe kann einen an manchen Tagen mit ihrer unaufhörlichen Eintönigkeit zermürben, an anderen Tagen lässt einen gerade dies unerwartete Schönheit entdecken: Wolkentürme im endlos blauen Himmel, die bis hinter den Horizont hinunterzusteigen scheinen, in eine ebenso weite Parallelwelt. Manchmal schaust du unterwegs einfach nur aus dem Fenster und fragst dich, ob diese Fahrt jemals enden kann. Kann es sein, dass Zeit konstant ihr Tempo wechselt, aber auch Gewicht und Masse? Morgens noch träge fließend, aber in sich leicht und flüchtig? Mittags dann eilig voranstürmend, dabei aber dicht und schwer? – Viele Kilometer weit, nahezu zwei Tage lang, sind wir am einstigen Ufer des Aral-Sees unterwegs, dessen ehemals 68.000 Quadratkilometer große Wasseroberfläche durch exzessive Wassernutzung für den Baumwollanbau während der Sowjet-Zeit auf heute nur noch rund 8.000 Quadratkilometer zurückgegangen ist. Aral, die Siedlung am früheren Nordufer des Sees, liegt heute über 30 Kilometer vom heutigen Ufer entfernt. Der See, ehemals so groß wie der deutsche Bundesstaat Bayern oder ein Drittel so groß wie Laos, ist heute größtenteils eine Salzwüste. Die Gespräche wirbeln zwischen den Autos hin und her, verebben dann wieder,

stop for weeks now, without any problems. Even the Safari classic is stubbornly pressing onward. The view over the bonnet of our travelling companions as they move ahead through the dust is an image we even see at night in our dreams.

At Baikonur, we catch a glimpse of the legendary Soviet space center, while around us another kind of space seems almost boundless. The Kazakh steppe can sometimes wear you down with its incessant monotony, while on other days it is precisely this monotony that makes you discover unexpected beauty: towering clouds in an endless blue sky that seem to come to earth just over the horizon as if from an equally vast parallel world. Sometimes you find yourself gazing out the window along the way and wondering if this trip will ever end. Is it possible that time can constantly change its pace, as well as its weight and mass, flowing sluggishly in the morning, but still light and fleeting? Rushing ahead at noon, but feeling dense and heavy? – For many kilometers, almost two days in fact, we follow the former shoreline of the Aral Sea. Formerly with a surface area of 68,000 square kilometers, the sea has shrunk to only about 8,000 square kilometers today due to excessive use of water in cotton cultivation during the Soviet era. Aral, the settlement on the former northern shore of the lake, is now more than 30 kilometers from today's shore. The lake, once the size of the German state of Bavaria or one third the size of Laos, is now largely a salt desert.

The chat passes back and forth between the cars, then ebbs away again, as everyone drives alone with their thoughts. Nonetheless, we have become an inseparable community of companions, the atmosphere is comradely, attentive and open. We have each other's backs – there is no other way to make such a trip. This is not just because of the need for organization, assistance and support, but above all because the grueling enormity of distances covered cries out to be shared with people who are experiencing exactly the same thing. You suspect that this is something that you won't really be able to explain at home. In the eyes of your fellow travelers you see the same drive, the same commitment, the same feeling of powerlessness, and the same determination that you feel.

When faced with the steppe landscape that stretches around us in all directions until it disappears into the blue haze of the vastness, it seems hard to imagine that this state of affairs will ever

jeder fährt ganz allein mit seinen Gedanken. Und trotzdem sind wir mittlerweile eine unzertrennliche Gemeinschaft von Gefährten, die Stimmung ist kameradschaftlich, aufmerksam, zugewandt. Wir haben einander im Blick – anders kann man eine solche Reise auch nicht machen. Und das nicht nur wegen Organisation, Beistand, Unterstützung, sondern vor allem, weil die zermürbende Ungeheuerlichkeit der Entfernungen geteilt werden will. Mit Menschen, die genau dasselbe erleben. Du ahnst, dass das zu Hause nicht wirklich erzählt werden kann. Aber in den Augen deiner Kameraden siehst du dieselbe Getriebenheit, dieselbe Ergebenheit, dieselbe Ohnmacht, dieselbe Entschlossenheit, wie sie in dir ist.

Angesichts von Steppen-Landschaften, die ringsum in alle Richtungen reichen, bis sie im blauen Dunst der Weite verschwinden, scheint es kaum vorstellbar, dass dieser Zustand jemals anders werden kann. Stundenlang bewegst du das Lenkrad keinen Millimeter, entdeckst sämtliche Schattierungen des Geradeausfahrens. Kasachstan geht trotzdem unter die Haut, mit großer Intensität. Der Seidenstraßen-Flow hat uns gepackt. – Unterbrechungen, Pausen und Zwischenstopps saugen wir trotzdem regelrecht auf, stellen dabei fest, dass sich mit den Kulturen und dem Land auch das Essen geändert hat. Hier kommt das auf den Tisch, was ein trockenes, extremes Land hergibt: Fleisch. Kasachstan röstet saftige Grillspieße auf Holzkohle-Feuern, sonnenhungriges Gemüse und Zwiebeln kommen dazu. Die Vieldimensionalität der chinesischen Küche und die Subtilität südostasiatischer Gerichte gibt es hier nicht. Aber irgendwie passt es zusammen, das Fahren und das Essen. Rustikal, nahrhaft und etwas zu Kauen. Endlos weit unter einem ewigen Himmel.

Erst in Aqtöbe drehen wir ab, von hier sind es nur noch rund 600 Kilometer bis nach Atyrau am Kaspischen Meer. Aus dem Nichts, dem weiten Raum, strömen wir heran, stehen plötzlich am größten Salzsee der Erde. Auf der anderen Seite warten Aserbaidschan, Armenien, Georgien – hier werden wir uns auf eine ganz andere Reise begeben: durch die Länder Europas. Wir drehen uns ein letztes Mal um, schauen zurück in die Weite, aus der wir gekommen sind. Eine Windhose saugt Staub empor, in einem mächtigen Strudel, der bis weit hinauf in dem Himmel ragt und dann zerfasert, in sich zusammenfällt. Sandkörner kitzeln uns im Gesicht. Auf Wiedersehen, Asien.

Sometimes you find yourself gazing out the window along the way and wondering if this trip will ever end. Is it possible that time can constantly change its pace, as well as its weight and mass, flowing sluggishly in the morning, but still light and fleeting? Rushing ahead at noon, but feeling dense and heavy? – For many kilometers, almost two days in fact, we follow the former shoreline of the Aral Sea. Formerly with a surface area of 68,000 square kilometers, the sea has shrunk to only about 8,000 square kilometers today due to excessive use of water in cotton cultivation during the Soviet era.

change. For hours, you don't move the steering wheel a single millimeter, discovering every possible shade of monotony. Still, Kazakhstan manages to get under your skin, with enormous intensity. We find ourselves in the grip of the flow of the Silk Road. Nevertheless, we welcome every interruption, break and stopover, realizing that the available food has changed in line with the cultures and the country. We discover what a dry terrain of extremes has to offer: meat. Kazakhstan's cuisine involves roasting juicy barbecue skewers on charcoal fires, with sun-hungry vegetables and onions as side dishes. The multidimensionality of Chinese cuisine and the subtlety of Southeast Asian dishes are unknown here. But somehow, the driving and the food work together. What we get is rustic and nutritious and something to chew on. Endless distances under an eternal sky.

We only turn off when we reach Aktobe. From here it is only about 600 kilometers to Atyrau on the Caspian Sea. Arriving out of nowhere, we find ourselves suddenly standing on the shore of the largest salt lake on earth. On the other side, Azerbaijan, Armenia, Georgia wait for us to embark on a completely different journey through the countries of Europe. We turn around one last time to look back into the vastness from which we came. An eddying wind sucks up dust in a mighty twist that rises far up into the sky, then frays and collapses. Grains of sand tickle our faces. Goodbye, Asia.

»Gute Reise« Party im Porsche Center Bangkok. Über 200 Freunde feiern mit uns den Aufbruch zu unserem Trip.
»Gute Reise« party at the Porsche Center Bangkok. More than 200 friends celebrate with us as we set off on our trip.

The Epic Bangkok - Stuttgart Porsche Drive 2025.
START

ลอตเตอรี่
ลอตเตอรี่
59

PORSCHE

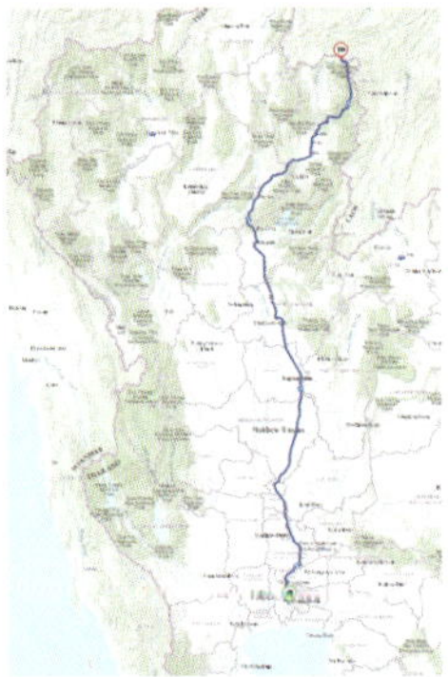

CURVISTAN
BO 992
THAILAND 01

The Epic Bangkok - Stuttgart
START
THE
EPIC
25
Reise!
PORSCHE
CURVISTAN
29 30

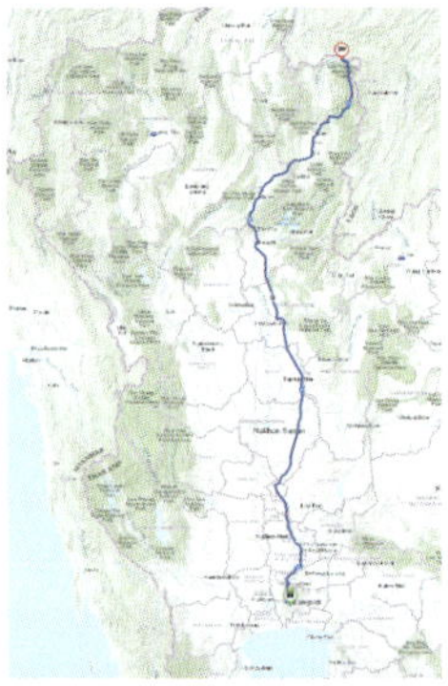

Auf geht's in den Norden. Wir grooven uns ein und freuen uns auf das große Abenteuer. Die Maschinen laufen, die Stimmung ist aufgekratzt. Off we go to the north. We're getting into the groove and looking forward to the big adventure. The machines are running, spirits are high.

ริมปิงวัสดุก่อสร้าง
CURVISTAN
LHD
TOW
4C 66
THAILAND 01

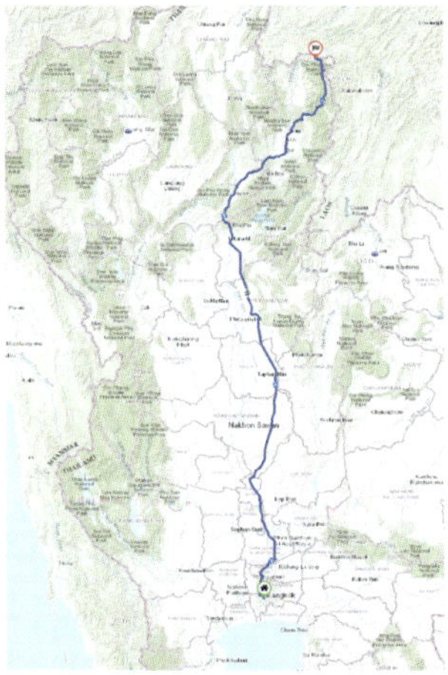

CURVISTAN
BO 992
THAILAND

22
22
CURVISTAN
AA 161
THAILAND

DAS TREFFEN
CURVISTAN
Leica
THE EPIC BANGKOK STUTTGART PORSCHE DRIVE
RENNDRIVE
MIDAS WINDOW FILM
TAG HEUER
BO 992
THAILAND

CURVISTAN
4C 66
THAILAND

CURVISTAN
BO 992
THAILAND

4C 66
THAILAND
PORSCHE

Die Kupplung »schmiert« – also Kupplungswechsel am Straßenrand mit Streetfood. Nach drei Stunden ging die Reise weiter. Beim 1986er Safari war das die einzige aufwendige Reparatur auf der gesamten Reise.
The clutch was slipping, so we changed it at the roadside – accompanied by some street food. After three hours, the journey continued. For the 1986 Safari, this was the only major repair during the whole trip.

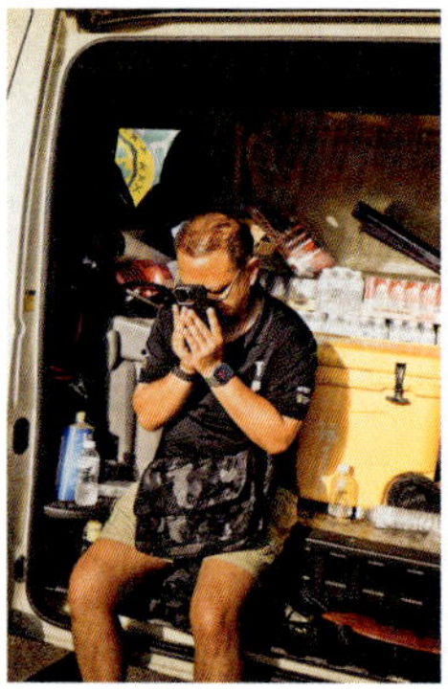

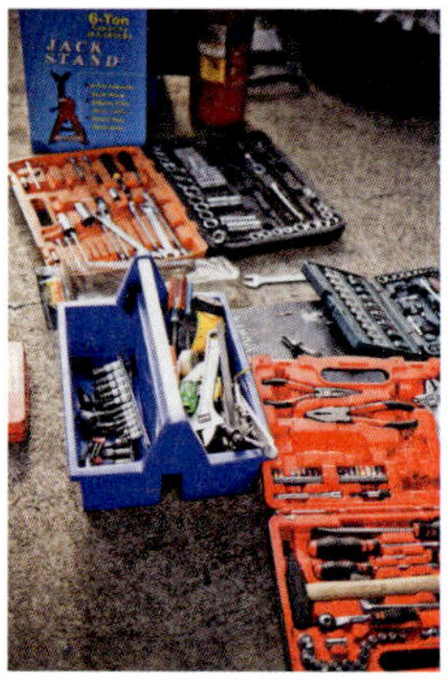

4C 66
THAILAND

SERVICE CREW

Abschied von Thailand - Grenzübertritt Laos. Kurze Wartezeit, Visas gecheckt, weiter geht die wilde Jagd. Farewell to Thailand – border crossing into Laos. A short wait, visas checked, and the wild ride continues.

CURVISTAN
PORSCHE
CURVISTAN
AA 161
THAILAND

Laos – etwas schwierige Straßenverhältnisse: Für 200 km haben wir zwei Tage benötigt. Dafür ist das Essen hervorragend. Laos – the road conditions are a bit challenging; it took us two days to cover 200 km. But the food is excellent.

CURVISTAN
4C 66
THAILAND 01

Einreise nach China – was wird uns erwarten? Kommen die Autos ohne Probleme über die Grenze? »Ake«, unser Mastermind, hat gefühlt 10 kg Papierkram dabei. Akribische Vorbereitung zahlt sich aus. Erster Kontaktpunkt an der Grenze: Der Computer zur Passkontrolle spricht Deutsch und führt durch die Formalitäten. Nach fünf Minuten – fertig. Wow. Entering China – what awaits us? Will the cars make it across the border without any issues? »Ake«, our mastermind, is carrying what feels like 10 kilos of paperwork. Meticulous preparation pays off. First point of contact at the border: the passport control computer speaks German and walks us through the formalities. Five minutes later – done. Wow.

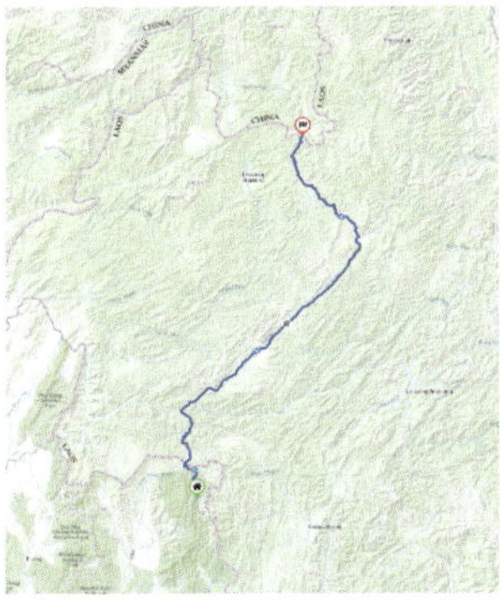

中国

29 30
THAILAND

尚勇二路 尚勇正街
纳勇大道

A04286

A04288

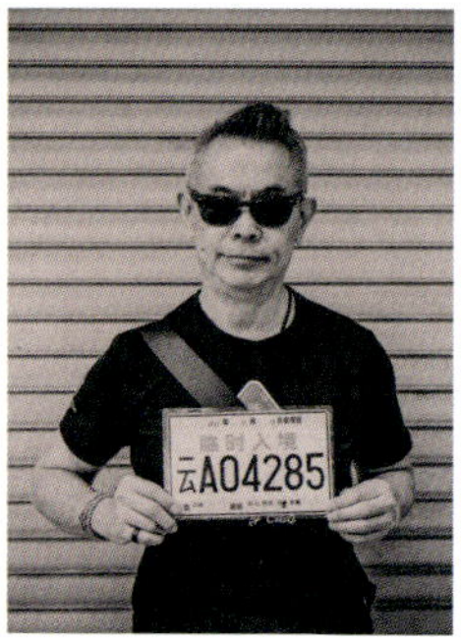
A04285

A04282

Willkommen in China. Die Einreise war schnell und unkompliziert. Für China benötigen wir eine kurze Einweisung bei der lokalen Führerscheinstelle und holen dort auch unsere Kennzeichen ab. Welcome to China. Entering the country was quick and straightforward. For driving in China, we need a brief orientation at the local licensing office and we pick up our license plates there.

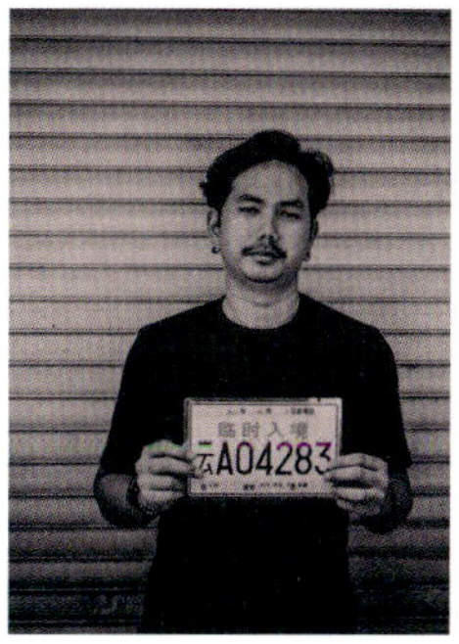

Erster Eindruck: Superfuture. Alles neu, alles sauber. Neben der Autobahn rauscht der Hochgeschwindigkeitszug vorbei. Jeder begrüßt und filmt uns auf unserem Weg und ruft uns euphorisch »Welcome to China« zu. Die Städte: historisch und zugleich supermodern.
First impression: super-future. Everything is new, everything is clean. Next to the highway, the high-speed train rushes past. Everyone greets and films us along the way, shouting an enthusiastic »Welcome to China.« The cities: historic and ultra-modern at the same time.

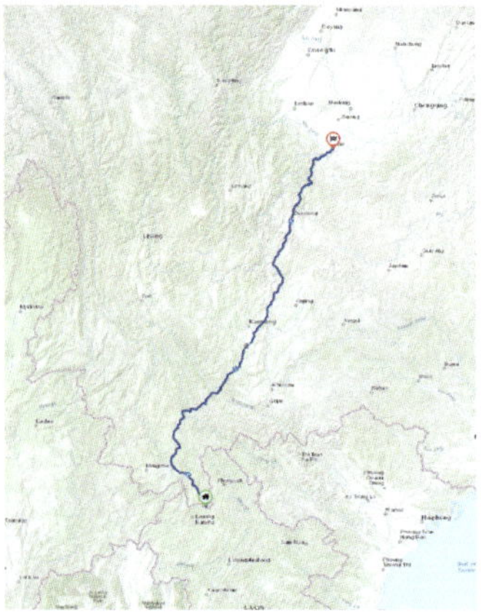

和美向未来
匠心酿品质

客车车道
车道关闭
军车免费通行
7座及以下小型客车免费通道
凌晨2时 至5时
短途客车
请走人工车道
20 Km/h
保持车距
绿色通道
专用车道
5 km/h
ETC人工
保持车距

宗范隧道

Die ersten Tage in China: Das Essen ist großartig, der Vibe in der Gruppe ebenfalls. Wir kommen wie geplant voran. The first days in China: The food is amazing, and the vibe in the group is great as well. We're making progress just as planned.

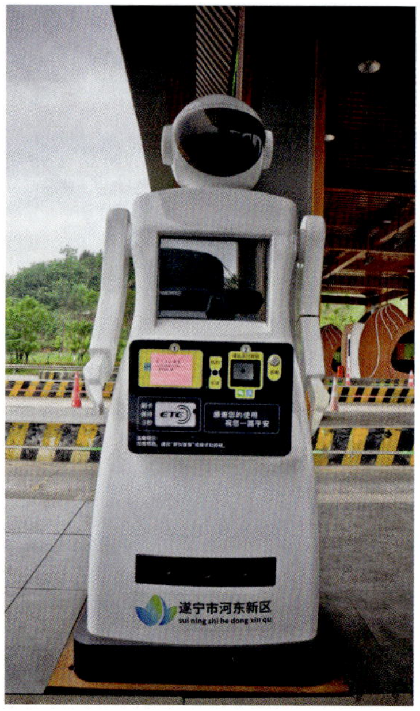

80

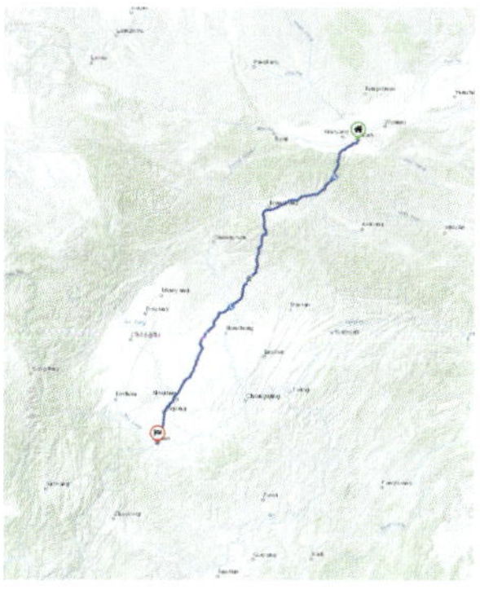

Auf dem Weg nach Xi'an. Wunderschöne Landschaft – das Wetter ist ein Traum. Die Food-Stops an den Tankstellen erinnern uns an italienisch Geräuchertes. In Xi'an werden wir vom Porsche Club herzlich begrüßt, und wir feiern bis spät in die Nacht. On the way to Xi'an. Beautiful landscapes – the weather is a dream. The food stops at the gas stations remind us of Italian cured meats. In Xi'an, we are warmly welcomed by the Porsche Club, and we celebrate late into the night.

Tausende Tonkrieger stehen in Reih und Glied – die Terrakottaarmee, seit über 2.000 Jahren Hüter des ersten Kaisers, noch immer ein Monument menschlicher Handwerkskunst. Gleich nebenan: Xi'an, uralte Kaiserstadt und Ausgangspunkt der Seidenstraße, von Marco Polo einst als pulsierender Knoten des Ostens beschrieben. Die Stadt lockt mit großartigem Streetfood – Biangbiang-Nudeln, dampfenden Baozi und den Düften des muslimischen Viertels. *Thousands of clay warriors stand in perfect formation – the Terracotta Army, guardians of the first emperor for over 2,000 years and still a breathtaking monument to human craftsmanship. Just beyond lies Xi'an, ancient imperial capital and gateway to the Silk Road, once described by Marco Polo as a vibrant hub of the East. Today the city draws you in with outstanding street food – Biangbiang noodles, steaming baozi, and the aromas of the Muslim Quarter.*

陕西交控

Porsche Club
China

共创美好未来

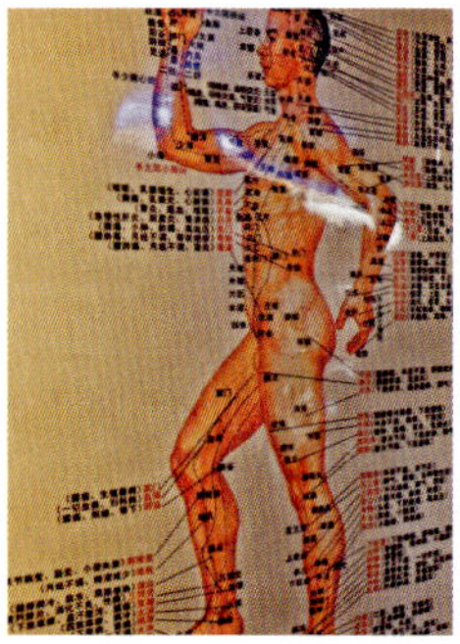

Nach tausenden Kilometern Richtung Norden geht es nun Richtung Osten. Die Landschaft verändert sich nur leicht, das Essen ist weiterhin großartig – jeder Tag ist ein Traum, an den man immer wieder gerne zurückdenkt. After thousands of kilometers heading north, we now turn east. The landscape changes only slightly, the food remains outstanding – every day is a dream you'll gladly return to in your memories.

CURVISTAN
BO 992
THAILAND

R
宽粉
水饺
面食

CURVISTAN
AA 161
THAILAND 01

Auf einmal ändert sich die Landschaft schlagartig. Steppe, Wüste und Gebirge. Aus Grün wird Braun – der Horizont wird endlos. Suddenly, the landscape changes dramatically. Steppe, desert, and mountains. Green turns to brown – the horizon stretches endlessly.

紫雲阁酒樓
明清街店
全球加盟热线
4000-6262-83
SINGHA
DAS TREFFEL
XPEL

七彩丹霞
unesco

积极行动 人人参与 全面创建全国文明旅游示范单位
抓学习促提升

Zwischen den farbigen Schichten der Erde entfalten die Rainbow Mountains im Zhangye Nationalpark ein fast unwirkliches Panorama – wie von Hand gemalt, nur eben von der Natur. Eine Landschaft die das Fernweh aufdreht. Between the earth's colored layers, the Rainbow Mountains of Zhangye National Park unfold a near-surreal panorama – as if painted by hand, only crafted by nature itself. A landscape that instantly turns up the wanderlust.

66
4C 66
THAILAND

BO 992
THAILAND

Auf der Fahrt weiter Richtung Westen geht einem als Alpenfan das Herz auf. Die Berge im Hintergrund sind 5000er, und der Drang hineinzufahren ist riesig. Das müssen wir aber verschieben … Heading further west, any Alps lover's heart instantly lifts. The mountains rising in the distance are 5,000-meter peaks, and the urge to drive up into them is overwhelming. But we'll have to save that for later…

Das Jiayuguan-Fort liegt malerisch in einem weiten Tal, nur wenige Kilometer westlich der Stadt Jiayuguan in der Provinz Gansu. Hier beginnt der westliche Abschnitt der Großen Mauer der Ming-Dynastie – ein Ort, der nicht ohne Grund den Titel »größter Pass unter dem Himmel« trägt. Seit seiner Errichtung im Jahr 1372 trotzt die Festung seit über 600 Jahren Wind, Wüste und Geschichte. Jiayuguan Fort sits dramatically in a broad valley just a few kilometers west of Jiayuguan City in Gansu Province. This is where the western end of the Ming-dynasty Great Wall begins – a place rightly known as the »Greatest Pass under Heaven.« Built in 1372, the fortress has stood strong for more than 600 years, weathering desert winds and the march of history.

华运连
服务
way service
正在营业
汽修 补胎 配件
冻液 轮胎批发
值班电话: 15032665777

汽修
高台店
配件维修
RENNDRIVE
SINGHA
LEMON SODA
DAS TREFFEN
XPEL
4C 66
THAILAND

RENNDRIVE
66

momo

Hier oben im Norden fahren wir an Windparks entlang, die ihresgleichen suchen. 120 km entlang eines Windparks mit 18.000 Windrädern – nur einer von vielen in China. Hier rechts im Bild ein Windradtransporter, von denen wir sicher über 40 Stück an einem Tag überholt haben.
Up here in the north, we drive past wind farms unlike anything we've ever seen. 120 kilometers along a wind farm with 18,000 wind turbines – just one of many in China. In the photo on the right you can see a wind turbine transporter – we must have passed well over 40 of them in a single day.

瓜
SINGHA
DAS TREFFEN
XPEL

Die Mingsha-Sanddünen bei Dunhuang sind für ihr »singendes« Geräusch bekannt, das entsteht, wenn der Wind über die Hänge streicht. Die Dünen gehören zur Kumtag-Wüste und erstrecken sich über rund 40 Kilometer Länge und 20 Kilometer Breite. Direkt am Rand der Sandberge liegt der Mondsichelsee – ein markanter Wasserpunkt inmitten der weitläufigen Landschaft.

The Mingsha Sand Dunes near Dunhuang are known for the »singing« sound created when wind moves across their slopes. Part of the Kumtag Desert, the dunes stretch roughly 40 kilometers east to west and 20 kilometers north to south. At their edge lies Crescent Lake – a distinct patch of water set against the wide, open desert landscape.

Einer meiner Lieblingstage in China – die Etappe war einfach der Wahnsinn: 700 km endloser Flow. Und am Ende des Tages ein Seafood-Restaurant im Nirgendwo. Es gab auf der Etappe zwar eine Tankstelle – aber die war rein elektrisch ausgelegt. Mit dem letzten Liter Sprit haben wir es dann doch geschafft. *One of my favorite days in China – the stage was absolutely insane: 700 kilometers of uninterrupted flow. And at the end of the day, a seafood restaurant in the middle of nowhere. There was a service station along the route – but it was designed exclusively for electric vehicles. In the end, we made it on the very last liter of fuel.*

五坛
青稞酒

Die Tempelanlage von Mogao hat eine bemerkenswerte Entstehungsgeschichte: Im Jahr 366 soll der Mönch Leig Tsung auf der Seidenstraße eine Vision gesehen haben – ein Licht, aus dem tausend Buddhas hervortraten. Tief beeindruckt zog er sich in die Oase zurück und begann, die erste Höhle in den Fels zu schlagen. In den folgenden Jahrhunderten folgten ihm zahlreiche Pilger und Mönche, sodass hunderte kunstvoll gestaltete Grotten entstanden – heute eines der bedeutendsten buddhistischen Heiligtümer der Welt.
The Mogao cave complex has an extraordinary origin story. In 366 AD, the monk Leig Tsung is said to have witnessed a vision of light and »a thousand Buddhas« while traveling along the Silk Road. Inspired, he retreated to the nearby oasis and carved the first cave into the cliff. Over the next centuries, pilgrims and monks followed his example, creating hundreds of beautifully decorated grottoes – now considered one of the world's most important Buddhist temple sites.

Altay
Taldy-Korgan
Almaty
Urumqi
Hami
Liuyuan
Lop Nur
Kashi
Hotan
TAKLA MAKAN
DESERT
KUNLUN MOUNTAINS

RENNDRIVE

固基创道
CURVISTAN
BO 992
THAILAND

CURVISTAN
BO 992

农夫山泉

陕汽重卡
430
15852333678

ZGYX 421T
SIT

Coffee

The most beatiful travel

29 30
THAILAND

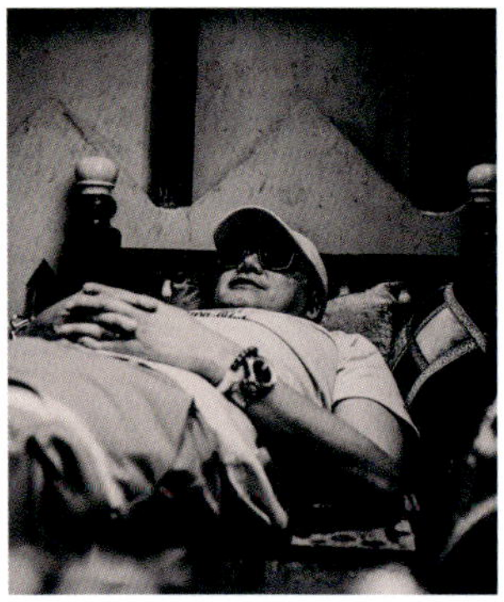

Urumqi ist die Hauptstadt der autonomen Region Xinjiang und gilt als eine der am weitesten vom Meer entfernten Großstädte der Welt. Die Stadt verbindet moderne chinesische Architektur mit zentralasiatischen Einflüssen, sichtbar vor allem im lebhaften International Grand Bazaar. Als Tor zum Tianshan-Gebirge ist Urumqi zugleich Ausgangspunkt für beeindruckende Berg- und Wüstenlandschaften rund um den Himmelssee.

Urumqi is the capital of the Xinjiang Autonomous Region and is considered one of the most landlocked major cities in the world. The city blends modern Chinese architecture with Central Asian influences, most evident in the lively International Grand Bazaar. As the gateway to the Tianshan Mountains, Urumqi is also the starting point for impressive mountain and desert landscapes surrounding Heavenly Lake.

疆小豹
博州消防救援支队

Jetzt wird es richtig spannend. Nach fünf Tagen wüstenähnlichem Cruisen erreichen wir die Grenze zu Kasachstan. Plötzlich erheben sich gewaltige Berge – Schnee trifft Pekingente. Ein Gefühlstornado. Oben auf dem Pass – das könnte auch der San-Bernardino-Pass sein – ist man auf einmal in Kasachstan gelandet. Wie in einem Paralleluniversum. Wüste, Seen, Berge und eine Weite, die ihres-gleichen sucht. Now it really gets exciting. After five days of desert-like cruising, we reach the border with Kazakhstan. Suddenly, towering mountains rise up – snow meets Peking duck. An emotional whirlwind. At the top of the pass – it could just as well be the San Bernardino – you suddenly find yourself in Kazakhstan. Like a parallel universe. Desert, lakes, mountains, and a vastness beyond compare.

SINGHA

果子沟

SINGHA

ABERFELDY
24
EXCEPTIONAL CASK

Goodbye China. Die beschleunigte Reise durch dieses riesige Land hat uns alle beeindruckt, und wir schmieden bereits Pläne, welche Route wir beim nächsten Mal nehmen könnten. China hat uns alle positiv in seinen Bann gezogen – Mindset und Perspektive verändern sich, wenn man es selbst erfährt. Und das ist das Schönste am Reisen: Es erweitert den Horizont. Goodbye China. The fast-paced journey through this vast country impressed all of us, and we're already making plans for which route we might take next time. China captivated us all in a very positive way – your mindset and perspective change when you experience it firsthand. And that's the most beautiful thing about traveling: it broadens your horizons.

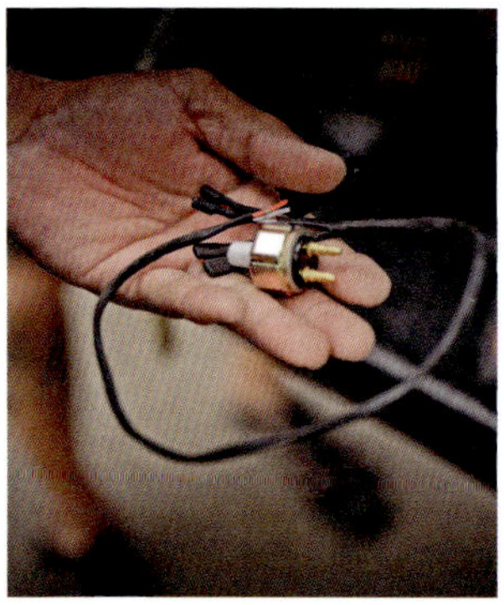

Almaty liegt am nördlichen Rand des Tian-Shan-Gebirges und verbindet urbane Dichte mit unmittelbarer Nähe zur Natur. Als ehemaliges kulturelles und wirtschaftliches Zentrum Kasachstans bleibt Almaty bis heute der wichtigste urbane Knotenpunkt des Landes. In den umliegenden Bergregionen finden sich die historischen Wildapfelwälder – Ursprung der Apfelsorte Malus sieversii und Grundlage für den Beinamen »Stadt des Apfels«.

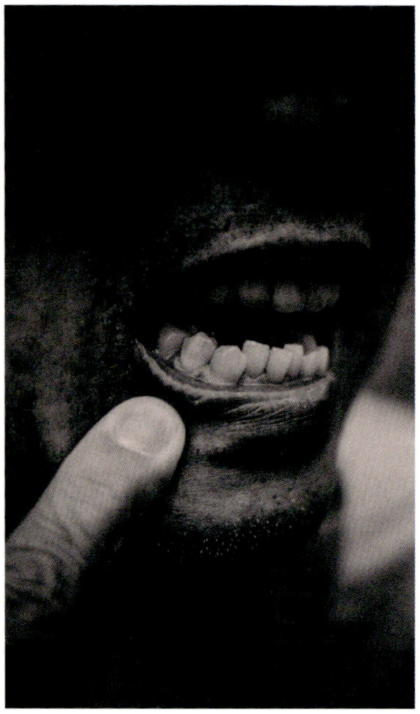

Almaty is located on the northern edge of the Tian Shan Mountains and combines urban density with immediate proximity to nature. As the former cultural and economic center of Kazakhstan, Almaty remains the country's most important urban hub to this day. In the surrounding mountain regions, you can find the historic wild apple forests – the origin of the apple species Malus sieversii and the basis for the city's nickname, the »City of the Apple.«

Für die nächsten sechs Tage geht es durch die Steppe Kasachstans. Wir staunen darüber, wie man solche Entfernungen vor Hunderten von Jahren überwinden konnte. In Kasachstan wird es im Winter bis zu minus 50 Grad kalt und im Sommer bis zu 50 Grad heiß. For the next six days, we will travel through the Kazakh steppe. We are amazed at how people were able to cover such distances hundreds of years ago. In Kazakhstan, temperatures can drop to minus 50 degrees in winter and rise to 50 degrees in summer.

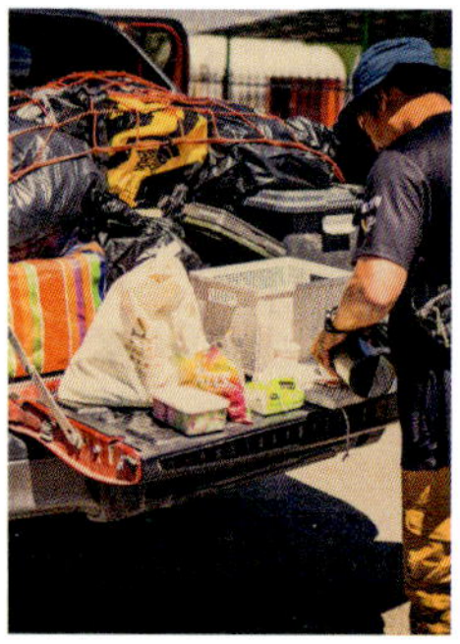

Heritage
Motoring Club

SINGHA
LEMON SODA
DAS TREFFEN
XPEL
4C 66
THAILAND

BO 992

SINGHA

KAZAKHSTAN
UZBEKISTAN

ПРОТОН-М

БАЙКОН
ПУТЬ К ЗВЕЗДАМ!

Ein Must-Visit-Stop: Der Weltraumbahnhof Baikonur in Kasachstan zieht wie ein geschwungener Korridor durch die Steppe und gilt als ältester sowie bedeutendster Raketenstartplatz der Welt – von hier erhoben sich einst Sputnik 1 und Juri Gagarins erster Flug in die Weite des Alls.
A must-visit stop: The Baikonur Cosmodrome in Kazakhstan stretches like a curved corridor through the steppe and is considered the oldest and most significant rocket launch site in the world – from here, Sputnik 1 and Yuri Gagarin's first courageous journey into the expanse of space once lifted off.

8
BANGKOK • STUTTGART

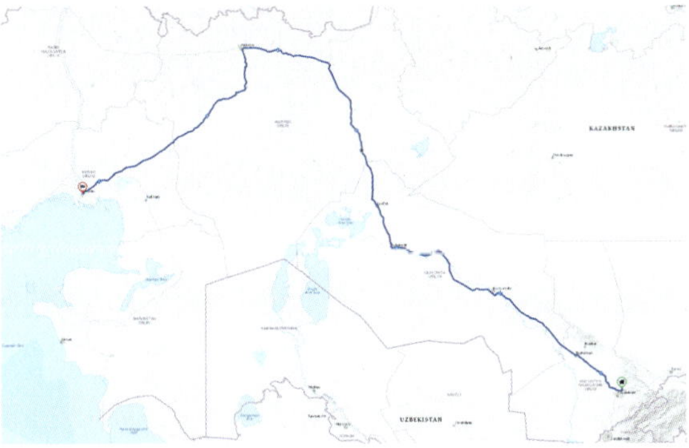
KAZAKHSTAN
UZBEKISTAN

Etappe zwei
Stage two

GEORGIEN / TÜRKEI / GRIECHENLAND / BULGARIEN / RUMÄNIEN UNGARN / SERBIEN / BOSNIEN & HERZEGOWINA / KROATIEN SLOWENIEN / ITALIEN / ÖSTERREICH / DEUTSCHLAND

GEORGIA / TURKEY / GREECE / BULGARIA / ROMANIA / HUNGARY SERBIA / BOSNIA & HERZEGOVINA / CROATIA / SLOVENIA / ITALY AUSTRIA / GERMANY

Die Gondel glänzt silbern im Sonnenlicht, wiegt auf den ersten Metern sanft am Kabel, schwebt dann ruhig hinunter in die Stadt. Hinter uns bleibt ein schroffer Berggrat zurück, auf dessen Gipfel die bulligen Mauertürme der Nariqala-Festung seit über 1.700 Jahren Front gegen imaginäre Angreifer machen. Georgien liegt eben bereits seit der Antike zwischen kriegerischen Nachbarn, wurde von den Römern an die Perser, von den Byzantinern an die Araber weitergereicht, schließlich vom russischen Kaiserreich geschluckt und dann in die Sowjetunion integriert – unabhängig ist der Kaukasus-Staat erst seit 1990. Kein Wunder, dass in der Hauptstadt Tiflis die Nationalfarben Weiß und Rot überall eingewoben sind. Das georgische Nationalbewusstsein ist uralt, Sprache und Schrift werden ebenso gepflegt wie Musik und Kultur.

Unten im Tal windet sich die Kura braun und träge dahin, Wohnhäuser stürmen tumultartig die steilen Hänge, Kirchtürme und Kathedralen ragen dazwischen auf. Die engen Kopfsteinpflaster-Gassen erzählen Geschichten aus Jahrhunderten, voller Gewürz- und Blütenduft, Vogelgezwitscher und Stimmengewirr. Auch die Küche Georgiens ist vom Land zwischen den Bergen und den Kulturen geprägt, zum Eckpfeiler ernennen wir den Chatschapuri-Käsefladen, der einer italienischen Focaccia oder Pizza ziemlich nahe kommt. Dazu gibt es eine Kharcho-Suppe mit Lammfleisch oder Mzwadi-Lammspieße, die uns verdächtig an das erinnern, was in Kasachstan und anderswo in Zentralasien als Schaschlik über Holzkohle geröstet wird. Die von Fernost bis Westeuropa in vielen unterschiedlichen Formen vorkommenden Teigtaschen nehmen in Georgien echtes XL-Format an – Chinkali heißen sie und man füllt sie mit gewürztem Hackfleisch, Pilzen oder Käse. Bevor man sich aber diesen Köstlichkeiten anvertraut, wird ein Phkali als Vorspeise genossen, eine leuchtend bunte Paste aus Spinat und Roter Beete mit Walnüssen und Knoblauch. Sehr herzhaft, ein denkwürdiger Charakter.

Ein modernes Tiflis gibt es natürlich auch, die prägenden Architekturgiganten am Kura-Ufer sind kaum zu übersehen. Als archaischer Steinblock trotzt die georgisch-orthodoxe Metechi-Kirche aus dem 13. Jahrhundert auf den Felswänden des Steilufers noch gegen das 21. Jahrhundert an, aber dann geht es nördlich des Europa-Platzes los: Wie eine Woge aus Glas wabert

The cable car shines like silver in the sunlight, swaying gently back and forth on the cable for the first few meters, before gliding serenely down into the city. A rugged mountain ridge recedes behind us, its summit crowned by the imposing towers of the Narikala Fortress, which has stood guard against imaginary invaders for over 1,700 years. Georgia has been situated between warring neighbors since antiquity, passing from Roman to Persian rule, from Byzantine emperors to Arab potentates, then finally swallowed up by the Russian Empire and then integrated into the Soviet Union. The Caucasus nation has only been independent since 1990, so it's hardly surprising that the national colors of white and red are on prominent display throughout the capital, Tbilisi. Georgian national consciousness stretches back a long way; the country's language and script are just as cherished as its music and culture.

Down in the valley, the brown waters of the Kura River wind their languid way between the steep slopes, with homes clinging tightly to the inclines and church steeples and cathedral towers rising up among them. Filled with the scent of spices and blossoms, birdsong, and the murmur of voices, the narrow-cobbled streets whisper tales of centuries past. Georgian cookery is also shaped by the land between the mountains and its diverse cultures. Khachapuri cheese flatbread is a cornerstone of Georgian cuisine. This is quite similar to an Italian focaccia or pizza and is often accompanied by kharcho soup with lamb or mtsvadi lamb skewers, which bear a striking resemblance to what is known as shashlik, grilled over charcoal in Kazakhstan and elsewhere in Central Asia. The dumplings, which appear in many different forms from the Far East to Western Europe, take on a truly gigantic size in Georgia, where they are called khinkali and are filled with seasoned ground meat, mushrooms, or cheese. Before indulging in these delicacies, we sample an appetizer of phkali, a brightly colored paste made from spinach and beetroot with walnuts and garlic. Very hearty and a truly memorable dish.

Of course, Tbilisi also has a modern part. The iconic architectural giants ranged along the Kura River are hard to miss. The 13th-century Georgian Orthodox Metekhi Church still defies the 21st century, protruding like an archaic block of stone from the cliffs of the vertiginous bank. Then, north of Europe Square,

die Friedensbrücke über den Fluss, dahinter wälzen sich die glänzenden Tuben des Rike-Zentrums im Park und einen Steinwurf entfernt breiten sich titanenhafte Beton-Seerosenblätter schützend über die Glasquader des Komplexes aus georgischer Nationalbank und Gerichtsgebäude. – Die uralte Stadt zwischen den Welten Asiens und Europas, Russland im Norden und dem Orient im Süden hat in den letzten Jahrzehnten gewaltig Fahrt aufgenommen, Georgien beansprucht wie schon seit langer Zeit seinen eigenen Platz. Und das gilt nicht nur politisch und kulturell, sondern auch geografisch: Im Norden ziehen sich die bis zu 5.600 Meter hohen Berge des Großen Kaukasus über 1.000 Kilometer weit vom Schwarzen bis ans Kaspische Meer und im Süden schiebt sich der Riegel des Kleinen Kaukasus, bis zu 3.700 Meter hoch vor die Länder im Süden und Südosten, Armenien und Aserbaidschan.

Wir haben tausende Kilometer in den Knochen, die Wochen in den Weiten der Steppen und Wüstengegenden des westlichen Chinas und Kasachstans stecken uns unter der Haut. Ein Umweg über die Gebirgspässe im Norden reizt uns daher jetzt umso mehr, raus in die Berge, durch grüne Täler mit Blick auf gleißende Schneegipfel. Und Kurven, endlich Kurven, die das traumwandlerische Fahren in der unendlichen Weite Zentralasiens verblassen lassen. Fiebrig hetzen wir los, das Wolfsrudel hat wieder Witterung aufgenommen. Bis kurz vor Chaschuri surfen wir noch die Schnellstraße im Tal der Kura, dann geht es in den Norden. Erste Hügelketten bringen den Groove, hoch nach Sachkhere und von hier bis zum Khikhata-Pass ins Rioni-Tal. Immer wieder schauen wir erstaunt unseren Händen zu, die plötzlich zu tun haben: lenken, schalten, nach endloser langer Zeit im Wachkoma der Geraden.

Bei Zageri scheint dagegen die Zivilisation eine erste Pause einzulegen, zusammen mit dem schottergrauen, eiskalten Zchenischali-Fluss fahren wir in die Berge. Kleine Weiler und Siedlungen ducken sich schicksalsergeben, stundenlang kämpfen wir uns voran. Auf dem Weg zu den Gletscher- und Eisriesen am Zagari-Pass wird die Straße immer wilder. Die kernigen Offroader unseres Support-Teams sind jetzt in ihrem Element, der blaue G-Mercedes und unser Crew-Land-Cruiser haben kein Problem mit Furten, Schlaglöchern, Schotterpassagen. Und auch die Porsche-Meute rund um den 911 Dakar zieht unaufhörlich durch

the landscape takes a dramatic turn: the Peace Bridge snakes across the river like a wave of glass, beyond it the gleaming tubes of the Rike Center undulate in the park, and a stone's throw away, titanic concrete water lily pads spread protectively over the glass cubes of the Georgian National Bank and Courthouse complex. Straddling the worlds of Asia and Europe, with Russia to the north and the Orient to the south, this ancient city has gained tremendous momentum in recent decades, and Georgia, as it has for a long time, is claiming its own place among nations. This applies not only politically and culturally, but also geographically: to the north, the mountains of the Greater Caucasus, reaching up to 5,600 meters in height, stretch for over 1,000 kilometers from the Black Sea to the Caspian, and in the south, the barrier of the Lesser Caucasus, reaching up to 3,700 meters in height, pushes against the borders of Armenia and Azerbaijan to the south and southeast.

We feel the thousands of kilometers already traveled in our bones; the weeks spent in the vast steppes and deserts of western China and Kazakhstan are etched into our very being. This makes a detour through the mountain passes in the north all the more appealing – an escape into the mountains, through green valleys with views of dazzling snow-capped peaks. And curves, of course. Finally we encounter curves that make the trancelike driving across the endless expanse of Central Asia fade into insignificance. We set off feverishly; the wolf pack has picked up the scent again. We cruise along the highway in the Kura Valley until just before Khashuri, then head north. The first ranges of hills offer plenty of groove, taking us up to Sachkhere and from there to the Khikhata Pass in the Rioni Valley. Again and again, we find ourselves astonished that our hands suddenly have work to do: steering, shifting gears, after what felt like a coma-like eternity on straight highways.

Civilization seems to abandon us entirely near Zageri. Following with the gravel-gray, icy-cold Tskhenischali River, we head into the mountains. Small hamlets and settlements cower in an attitude of resignation as we struggle onward for hours. The road becomes increasingly wild on the way to the glaciers and ice giants at the Zageri Pass. The rugged off-roaders of our support team are now in their element; the blue G-Class Mercedes and our crew Land Cruiser have no problem handling fords, potholes,

Die Türkei ist beeindruckend, rau und süß zugleich, wie eine mit Walnuss gefüllte Dattel. Die ländlichen Gebiete strahlen träge Stetigkeit aus, selbst die Traktoren und Pickups, die TV-Antennen auf Hausdächern und Internet-Schaltkästen an Straßenecken können das Gefühl nicht abschütteln, dass das Leben hier seit Hunderten von Jahren seinen bewährten Gang geht, im Rhythmus von Saat und Ernte, Sonne und Regen, Sommer und Winter. Wilde Berge, karge Natur – dazu kommt uralte Geschichte, die bis in die Bronzezeit und noch weiter hinaus reicht.

– bis auf einen: Der vipergrüne GTS will mit ganzem Zartgefühl bewegt werden. Tastet sich immer wieder im Schritttempo voran, um den großen Bogen durch das Gebirge bis zum Enguri-Stausee zu schaffen.

Und dann sind wir da. Lassen die Berge hinter uns zurück, die 45.000-Einwohner-Stadt Sugdidi ist nach den Stunden im Großen Kaukasus beinahe ein Schock, bei Poti haben wir tatsächlich das Schwarze Meer erreicht. Klimawechsel. Landschaftswechsel. Stimmungswechsel: Nach dem Vorwärtstasten im Outback machen wir jetzt energisch Tempo und Strecke, bis Batumi an der Grenze zur Türkei. Nun haben wir sie, die ersten Kilometer am Ufer des Schwarzen Meers, und die Feststellung, dass nicht das Meer schwarz ist, sondern die Küste: Dunkle Hänge stürzen sich steil ins Wasser hinunter, sind dicht mit Gestrüpp bewachsen. Über der weiten Wasseroberfläche sammeln sich währenddessen düstere Gewitterwolken in allen Schattierungen von Lila über Dunkelgrau bis Blauschwarz – vielleicht kommt der Name auch daher?

Gespannt fahren wir, haben irgendwann tatsächlich die Grenzanlage zwischen Georgien und der Türkei erreicht. Dass wir bereits jetzt schon wieder an einer Staatengrenze stehen, kommt uns nach den vielen Tagen in Zentralasien seltsam vor – vermutlich werden wir uns aber an die Einreise- und Zoll-Stopps gewöhnen müssen: Auch wenn uns die Türkei noch ein paar Tage Schonfrist gewährt, stehen danach Griechenland und die Balkan-Länder auf dem Routenplan, und das bedeutet vermutlich unerfreulich oft: »Guten Tag, haben Sie etwas zu verzollen?« – Wir werden es mit Gelassenheit nehmen, auf einer 20.000-Kilometer-Reise ist jegliche Hetze ohnehin überflüssig. Ob diese großzügige Einstellung zum Lauf der Zeit auch wirklich ernst

or gravel sections. The Porsche pack gathered around the 911 Dakar continues relentlessly – all except for one: the viper-green GTS demands the utmost care from its driver. It cautiously inches its way forward at a snail's pace to make the wide arc through the mountains to the Enguri Reservoir.

Then, suddenly, we've arrived, having left the mountains behind. The city of Zugdidi, with its 45,000 inhabitants, is quite a shock after the hours spent in the Greater Caucasus. Near Poti, we've actually reached the Black Sea. We encounter a change of climate, a change of landscape, and a change of mood: after feeling our way forward in the outback, we now find ourselves energetically picking up speed and covering ground, all the way to Batumi on the border with Türkiye. We finally get to experience the first few kilometers along the Black Sea coast, realizing that it's not the sea that's black, but the coastline: steep, dark slopes densely overgrown with scrub plunge dramatically into the water. Meanwhile, gloomy storm clouds are gathering above the vast expanse of water in every possible shade from purple to dark gray to bluish-black – maybe that's where the name comes from? We drive on, full of anticipation, eventually reaching the border crossing between Georgia and Türkiye. It feels strange to come upon another national border so soon after so many days in Central Asia – but we realize we should probably get used to border controls and customs stops: even though Türkiye would grant us a few more days' respite, Greece and the Balkan countries are next on the itinerary, which would likely mean an unpleasantly frequent repetition of the words: »Good morning. Anything to declare?« – We decide to take it all in our stride; what's the hurry anyway when you're on a 20,000-kilometer journey?

»Nin« promptly tests whether this relaxed attitude to the passage of time is just skin deep by getting his Safari 911 stuck in deep sand during a short rally on a Black Sea beach. After being dug out by the team, he naturally has to ask the support crew to carry out a brake check to ensure that his youngtimer continues to perform equally well for the next few kilometers, despite being driven so uncompromisingly in the current harsh conditions. This beggars belief, as other cars of this age are generally allowed to accompany the emotional moments in their owners' lives with relative ease. Nin, however, goes hell for leather in his »Safari«, demanding just as much from his car as his fellow travelers. At this point, we have a particularly tough leg of the journey ahead of us: we head away from the Black Sea and out into the high-

gemeint ist, testet »Nin« prompt, er fährt seinen Safari-911 bei einer kleinen Rallye-Einlage am Schwarzmeer-Strand im tiefen Sand fest – und muss nach dem gemeinschaftlichen Ausgegrabenwerden selbstverständlich noch die Begleitmannschaft um einen Bremsencheck bitten. Damit der ungeniert im rauen Jetzt bewegte Youngtimer auch auf den nächsten Kilometern so gut läuft wie bisher. Kaum zu glauben: Andere seines Jahrgangs dürfen ihre Besitzer gut geschont durch gefühlvolle Lebensmomente begleiten, aber Nin schenkt seinem »Safari« nichts. Genau wie wir alle unsere Maschinen fordern. Und gerade jetzt haben wir eine ganz besonders grantige Etappe vor uns: weg vom Schwarzen Meer, hinaus ins Hochland Anatoliens. Das heißt: weit und schnell, kurvig und staubig. Eine wahre Sonderprüfung für Mensch und Material. Wir fahren hungrig und hechelnd, geradewegs durchs Innere der Türkei, landen bei Şanlıurfa und Gaziantep beinahe in Reichweite der syrischen Grenze und kurz vor Mersin sogar unweit des Mittelmeers.

Die Türkei ist beeindruckend, rau und süß zugleich, wie eine mit Walnuss gefüllte Dattel. Die ländlichen Gebiete strahlen träge Stetigkeit aus, selbst die Traktoren und Pickups, die TV-Antennen auf Hausdächern und Internet-Schaltkästen an Straßenecken können das Gefühl nicht abschütteln, dass das Leben hier seit Hunderten von Jahren seinen bewährten Gang geht, im Rhythmus von Saat und Ernte, Sonne und Regen, Sommer und Winter. Wilde Berge, karge Natur – dazu kommt uralte Geschichte, die bis in die Bronzezeit und noch weiter hinaus reicht. Am Göbekli Tepe bei bei Şanlıurfa werden gerade über 12.000 Jahre alte Bauten der Jungsteinzeit ausgegraben, aber das ist nur der spektakulärste Teil. Alle haben in diesem Land ihre Spuren hinterlassen, die Hurriter, Sumerer und Hethiter, die Seleukiden, Römer, Perser, Seldschuken und Mongolen – bis dann das aufsteigende Osmanische Reich zum bestimmenden Faktor Kleinasiens, rund ums Mittelmeer und hinauf zum Balkan wurde. Durch dieses Universum fahren wir, streben dabei immer weiter in Richtung Europa, im äußersten Westen der Türkei am Marmarameer.

Ein Erlebnis ist uns vor dem Wechsel zwischen den Kontinenten aber noch wichtig, wir haben spektakuläre Bilder im Kopf: Mächtige Tuff-Kegel in einer hochgelegenen Steppe zwischen Gebirgszügen, und darüber ganze Schwärme von bunten Heißluftballons. – Wir drehen daher von unserer Route in Richtung Westen ab, fahren stattdessen nach Norden, rund 240 Kilometer

lands of Anatolia. That means long distances and fast, winding and dusty conditions – a truly special test for both man and machine. Hungry and thirsty, we drive straight through the interior of Türkiye, arriving near Şanlıurfa and Gaziantep almost within reach of the Syrian border and just before Mersin, not that far from the Mediterranean.

Türkiye is a country that makes a big impression, both rugged and sweet at the same time, like a date filled with crushed walnuts. The rural areas exude a languorous steadiness; even the tractors and pickup trucks, the TV antennas on rooftops, and the internet junction boxes on street corners can't shake the feeling that life here has followed the same tried-and-tested pace for hundreds of years, following a rhythm of sowing followed by reaping, sunshine followed by rainfall, summer followed by winter. Wild mountains and merciless nature are allied with ancient history, stretching back to the Bronze Age and even further. Neolithic structures over 12,000 years old are currently being excavated at Göbekli Tepe near Şanlıurfa, but that's just the most spectacular part. Everyone has left their mark on this land: the Hurrians, Sumerians, and Hittites, the Seleucids, Romans, Persians, Seljuks, and Mongols – until the burgeoning Ottoman Empire became the dominant force in Asia Minor, around the Mediterranean, and up as far as the Balkans. We cruise through this universe, pressing ever closer to Europe, arriving at the Sea of Marmara in the far west of Türkiye.

One experience is still paramount for us before we quit one continent for another. We have spectacular images in our minds: a host of colorful hot air balloons floating above mighty cones of volcanic tufa rock in a high-altitude steppe landscape between mountain ranges. We turn away from our westward route and instead drive north for about 240 kilometers, over the Cilician Gates, a pass in the Taurus Mountains, and beyond into the highlands of Anatolia. Our destination is Göreme, a small town in Cappadocia.

We need to rise early the next morning to be driven over dirt tracks into a bizarre, dusty landscape that is barely recognizable in the semi-darkness. The silent, conical silhouettes of the typical rock towers of Cappadocia rise around us as the night sky slowly turns purple. We climb into the large basket of a balloon, pressing ourselves against the railing and watching intently as the teams prepare for take-off. We wait as a meter-long, hissing

weit, über die Kilikische Pforte, einen Pass im Taurusgebirge, und dahinter ins Hochland Anatoliens hinein. Das Ziel heißt Göreme, eine Kleinstadt in Kappadokien.

Am nächsten Morgen müssen wir früh aufstehen, werden über Feldwege in eine bizarre, staubige Landschaft hineingefahren, die in Halbdunkel kaum zu erkennen ist. Die stummen Kegel der typischen Felstürme Kappadokiens ragen wie Scherenschnitte rings um uns in einen sich langsam lila färbenden Nachthimmel. Wir steigen in die große Gondel eines Ballons, drängen uns gegen die Reling und schauen gespannt zu, wie die Teams das Gefährt startbereit machen. Schauen zu, wie eine meterlange, fauchende Flamme ins Innere der zarten Hülle fegt und diese langsam aufbläht. Plötzlich hebt sie vom Boden ab, schlingert noch schlapp und träge aufwärts, ruckt dabei unsere Gondel an, versetzt sie ein Stück weit. Wir halten uns krampfhaft fest, irgendwie kommt jetzt erst an, was da gleich mit uns geschehen wird! – Hochhausgroß ragt der Ballon über unseren Köpfen in den Himmel, die Gasflamme steht wie ein Fanal in seinem Inneren und lässt die Hülle leuchten. Ringsum ist im Zwielicht des Morgens dieses Leuchten zu sehen, überall erheben sich Ballons aus dem dunklen Boden, wachsen empor wie titanenhafte, magisch leuchtende Pilze. Und dann ist es so weit: Mit einem unsicher schlingernden Gefühl verlässt unsere Gondel den Grund. Die Haltemannschaft am Boden führt ihre Stabilisierungstaue nur noch ganz kurz, lässt dann los. Wir schweben. Wie in Zeitlupe scheint der Ballon an Höhe zu gewinnen – erst beim Blick nach unten wird uns klar, wie hoch wir bereits sind. Die Männer am Boden sind nur nach fingernagelgroß, schauen uns mit in den Nacken gelegten Köpfen nach. Rundum sackt das Land in die Tiefe, mitsamt den Schatten der Nacht – und wird genau in diesem Moment von der Morgensonne getroffen. Dunkelrot leuchtend kriecht der Sonnenball über den Horizont, taucht die Welt ins Licht. Wir stehen und staunen sprachlos. Schauen in die Runde, sind Teil eines Bilds in unserem Kopf geworden: Ein Ballon-Schwarm am Morgenhimmel. Lila, Orange, Blau, wunderschön. Und unter uns ein uraltes Land, das still hinaufsieht.
800 Kilometer weiter westlich haben wir einen Zehnstunden-Stint später das Mittelmeer südlich von Izmir erreicht, die Ägäis. Das Land des antiken Weltenbummlers Odysseus. Und weiter im Norden liegt Mazedonien, die Heimat Alexanders des Großen. Wir lächeln bei dem Gedanken – als Reisende von der Seidenstraße kommend, haben wir größere Distanzen bewältigt als diese Helden der Antike Aber gut, Alexander hatte keinen

flame sweeps into the interior of the delicate envelope and slowly inflates it. Suddenly the balloon lifts off the ground, lurches upwards still limp and sluggishly, pulling our basket in the process, moving it slightly. We hold on desperately. Somehow the reality of what is about to happen only hits us now! The balloon rises into the sky above our heads with the gas flame standing like a beacon inside it, illuminating the envelope, now as tall as a high-rise building. The same glow can be seen all around us in the twilight of the morning, as balloons rise from the dark ground, growing like gigantic, magically glowing mushrooms. Then the moment arrives: with an uncertain lurch, our gondola leaves terra firma. The teams on the ground only hold the tethers for a very short time before letting go. We are floating. As if in slow motion, the balloon seems to gain altitude. It is only when we look down that we realize just how high we have already climbed. The men on the ground are only the size of fingernails, watching after us with their heads tilted back. All around, the ground drops away, along with the shadows of the night – and we are hit by the rays of the morning sun at exactly this moment. Glowing dark red, the ball of the sun creeps over the horizon, bathing the world in light. We stand in speechless awe. We look around and find we have become part of an image in our heads: a swarm of balloons in the morning sky. Purple, orange, blue, beautiful. Below us lies an ancient land that gazes up at us in silence.

800 kilometers further west and a ten-hour drive later, we reach the Mediterranean south of Izmir, the Aegean Sea. This is the land of the Odysseus, the original ancient mariner. Further north is Macedonia, home of Alexander the Great. We smile at the thought that, as travelers coming from the Silk Road, we have covered greater distances than these heroes of antiquity.Admittedly, unlike us, Alexander didn't have a Porsche 911 Dakar, or Cayenne Hybrid to travel in. These trusty machines now whisk us along the coast to Çanakkale, the narrow inlet of the Dardanelles, through which the Mediterranean creeps into the Sea of Marmara. Across the waves in Istanbul is the mouth of the Black Sea, which we won't see again on this trip. We leave Asia behind high up on the Çanakkale Bridge, as Europe begins on the west bank of the Dardanelle straits. Exactly 125 kilometers later, we reach the border to Greece, or »Hellas«, as the Greeks themselves call their homeland. We almost feel guilty at the border crossing, because we'll be spending so little time here: we push north, driving just over 110 kilometers to the border with Bulgaria. You never know, maybe some day in the future we'll return for a visit

Porsche 911 Dakar, keinen Cayenne Hybrid – wir schon. Und diese Maschinen bringen uns nun klaglos entlang des Meers bis nach Çanakkale, zum schmalen Meeresarm der Dardanellen, durch den sich das Mittelmeer ins Marmarameer schleicht. Drüben in Istanbul mündet das Schwarze Meer, wir sehen es auf dieser Reise also nicht wieder. Hoch oben auf der Çanakkale-Brücke verlassen wir Asien, am Westufer des Dardanellen-Durchlasses beginnt Europa. Exakt 125 Kilometer später haben wir auch die Grenze nach Griechenland erreicht, oder »Hellas«, wie die Griechen selbst ihre Heimat nennen. Am Grenzübergang fühlen wir uns beinahe etwas schuldig, denn die Einreise lohnt sich für uns kaum: Wir streben nach Norden, fahren gerade einmal etwas über 110 Kilometer bis zur Grenze nach Bulgarien. Wer weiß, vielleicht kehren wir irgendwann in der Zukunft zurück, für einen Besuch in Athen, Sparta oder Thessaloniki. Aber nicht heute.

Immerhin stellen wir fest, dass Griechenland uns die verschämte Stippvisite nicht allzu schwer macht – oder besser: Bulgarien seit 2025 zum europäischen Schengen-Raum gehört, was die Grenzkontrollen zwischen den Staaten äußerst überschaubar hält. Gut ausgebaute Straßen kurven nach Norden, Bulgarien zeigt sich gute 370 Kilometer weit von seiner sonnigen Seite: Unterhaltsames, raumgreifendes Fahren in einem idyllischen Land, das selten langweilig wird. Und an der Donau, dem Grenzfluss zu Rumänien, müssen wir Bulgarien auch schon wieder verlassen: Der mächtige Strom durchquert weite Teile Europas, entspringt als kleine Quelle im südlichen Deutschland und mündet nicht weit entfernt von hier ins Schwarze Meer. Wir schauen den Fluss auf der Grenzbrücke genau an, er wird uns auf der weiteren Reise immer wieder begegnen. Donau. Danube. Duna. Wie auch immer er in den Ländern genannt wird, die er durchquert. Ein Reisender wie wir. Seelenverwandter.

Bis zur rumänischen Hauptstadt Bukarest sind es nur rund 70 Kilometer und von hier aus weitere 175 zum Fuß der Transsilvanischen Alpen – das lässige Meilenfresser-Tempo, mit dem wir bis hierher gekommen sind, darf nun wieder etwas angekurbelt werden: Auf der »Drumul Transfagarasan«, oder »Transfogarascher Höhenstraße« surfen wir über die Berge nach Norden. Die Transfagarsan ist ein absolutes Kurvenmonster, frisst und schlingt, hetzt bis auf über 2.000 Meter hinauf und fährt uns schwindelig. Könnte das eine erste Erscheinung der tatsächlichen Alpen sein? – Nun, nicht ganz. Zwischen den Bergen Transsilva-

Suddenly the balloon lifts off the ground, lurches upwards still limp and sluggishly, pulling our basket in the process, moving it slightly. We hold on desperately. Somehow the reality of what is about to happen only hits us now! The balloon rises into the sky above our heads with the gas flame standing like a beacon inside it, illuminating the envelope, now as tall as a high-rise building. The same glow can be seen all around us in the twilight of the morning, as balloons rise from the dark ground, growing like gigantic, magically glowing mushrooms.

---

to Athens, Sparta or Thessaloniki. But not today. Nonetheless, we recognize that Greece does not make our shame-faced flying visit too difficult. Because Bulgaria has been part of the European Schengen area since 2025, border controls between the two states are extremely smooth. Well-developed roads curve north and Bulgaria shows its sunny side for a good 370 kilometers: we enjoy entertaining, expansive driving in idyllic countryside that rarely gets boring. When we reach the Danube, the river that marks the border with Romania, we have to leave Bulgaria again. The mighty Danube traverses large swathes of Europe, rising as a tiny trickle in southern Germany and flowing into the Black Sea not far from here. We take a closer look at the river from the vantage point of the border bridge, knowing that we will encounter it again and again on the rest of our journey. Danube. Donau. Duna. Whatever it is called in the countries it crosses, it is a constant soulmate for travelers like ourselves.

It is only a distance of about 70 kilometers from here to Romania's capital Bucharest and then another 175 kilometers to the foothills of the Transylvanian Alps. We now have an opportunity to crank up the casual cruising speed that has taken us this far. Taking the »Drumul Transfagarasan«, or »Transfagarasan Highway«, we surf over the mountains to the north. The Transfagarasan features endless, all-consuming curves, rushing us to altitudes of over 2,000 meters and making our heads spin. Could this be the first appearance of the actual Alps? Not quite. Between the mountains of Transylvania and the Alps of Central Europe lies the immense Pannonian Plain, a sandy basin hundreds of kilometers wide. While it doesn't compare to the wastes of Kazakhstan, here, too, every kilometer demands a persistently calm driving style. That's why we decide to descend into the valley. 400 kilometers later we are already in the middle of the sandy monotony

niens und den Alpen Mitteleuropas liegt die immense Fläche der pannonischen Tiefebene, einem sandigen, Hunderte Kilometer weiten Kessel. Kein Vergleich zu Kasachstan, aber auch hier will jeder Kilometer gefahren sein, hartnäckig, gelassen, selbstverständlich. Und deshalb runter ins Tal, 400 Kilometer später sind wir an der Grenze Rumäniens nach Ungarn bereits mittendrin im sandigen Monothema Pannoniens.

Die weitere Route soll uns noch einmal ans Mittelmeer führen, drüben in Kroatien, aber vorher wollen wir noch die Hauptstadt Ungarns besuchen: Budapest. Die Stadt liegt knappe 170 Kilometer weiter nördlich von unserer Reise-Route, ist aus dem Süden über eine Autobahn gut zu erreichen – und wer aus Bangkok bis hierher gekommen ist, lässt sich von einem 340-Kilometer-Umweg nicht abschrecken. Also: Budapest. Oder besser: Buda und Pest. Die Stadt besteht aus zwei Teilen, getrennt von der Donau, und zeigt sich bei unserem Besuch von der besten Seite: Türme und Zinnen, die Bastei-Festung des Burgviertels auf dem Westufer der Donau, majestätische Wolken über der Ebene und die Silhouetten von rollenden Hügelketten in weiter Ferne.

Zusammen mit der Donau fahren wir danach wieder in Richtung Süden, während sich ringsum die ungarische Puszta in alle Himmelsrichtungen zum Horizont zieht. Staubige, sandige Weite. Kleine Weiler im Schatten von Silberpappeln, hinter den einstöckigen Häusern werden Paprika und Tomaten gezogen. In Kasachstan haben wir gelernt, dass der Eindruck von Einöde täuschen kann, wenn man lange genug hinsieht, beginnt eine subtile Vielfalt zu sprechen. Alles in den Ebenen scheint einem regelmäßigen Rhythmus zu folgen, hinter dessen hypnotischem Puls eine meditative Kraft steckt. Fahren wie Bären Winterschlaf halten: Herzschlag einmal pro Minute, Stoffwechsel auf Standby – so vergehen Stunden. Einen gefühlten Augenblick später landen wir so in Serbien, fahren vorbei an der Hauptstadt Belgrad und stellen zuerst nur in Nuancen, dann immer deutlicher fest, dass der Steppen-Bann der pannonischen Tiefebene langsam nachlässt. Auftauen, Puls beschleunigen, Sitzposition zurechtrücken, vielleicht sogar das Lenkrad heranziehen. – Spätestens hinter dem Fluss Save ist das Land schließlich wie ausgewechselt, kleinteilig, hügelig, wildromantisch. Knorrige Straßen führen

of Pannonia at the border between Romania and Hungary. The remaining route should take us to the Mediterranean, across in Croatia, but before that we plan to visit Budapest, the capital of Hungary. The city is just under 170 kilometers further north of our travel route and is easily accessible from the south via a highway. Having come this far from Bangkok, a 340-kilometer detour is no deterrent. So… Budapest, here we come. Or rather: Buda and Pest. The city is in two parts, divided by the Danube, and puts on a spectacular show for our visit: towers and battlements, the bastion of the Castle District on the west bank of the river, majestic clouds over the plain and the silhouettes of rolling hills in the far distance.

Sticking close to the course of the Danube, we then head south again, the dusty, sandy expanse of the Hungarian Puszta stretching to the horizon around us in every direction. Small hamlets cluster in the shade of silver poplars, peppers and tomatoes growing behind the single-storey houses. While driving in Kazakhstan, we learned that the impression of wasteland can be deceiving and, if you look long enough, a subtle diversity becomes evident. Every element in the plains seems to follow a regular rhythm, with a hypnotic pulse that assumes a meditative power. Like bears in hibernation, our heartbeat seems to slow to once a minute, our metabolism having entered standby mode. Hours pass. In what feels like just a moment later, we arrive in Serbia, driving past the capital Belgrade. At first it is barely noticeable, then it becomes increasingly clear that the spell of the Pannonian Plain is slowly fading to become a memory. Time to rouse ourselves from slumber, get the blood flowing again, adjust our seats and maybe even move the steering wheel. After we cross the Sava River, there is no longer any doubt: the landscape has changed completely into something small-scale, hilly, and wildly romantic.

Gnarly roads take us over the mountains to Bosnia-Herzegovina, a country whose troubled past is part of recent European history. East meets West here in a crash of contradictory ideologies, each vying for influence. Gazing at the countryside between Sarajevo and Mostar, it is hard to believe that this has been the site of so much conflict with so little peace. For us, however, the summer

über die Berge nach Bosnien-Herzegowina, ein Land, dessen unruhige Geschichte Teil jüngerer europäischer Entwicklungen ist. Ost und West branden hier gegeneinander, auf- und absteigende Einfluss-Sphären. Das Land zwischen Sarajevo und Mostar lässt es nicht vermuten, aber ruhig war es hier nie. Nicht übermäßig. Nur für uns machen die sommerlichen Berge mit den kühlen Flusstälern eine Ausnahme, lassen die Vögel zwitschern, Eidechsen über alte Steinmauern kriechen, das Leben herrlich leicht dahinziehen. Unter der »Stari Most«, der alten Brücke Mostars strudelt eiskalt und grün die Neretva, wir schauen hinunter ins Wasser und fühlen der Schwerelosigkeit dieses Tages nach.

Kroatien und das Mittelmeer sind nun nicht mehr weit von hier entfernt, bereits einen gefühlten Augenblick später sind wir an der schroffen Steilküste unterwegs, die uns nun auf eine weit über 500 Kilometer reichende Reise nach Norden mitnimmt. Adria heißt dieser Teil des Mittelmeers, auf der anderen Seite liegt Italien und wir fahren hier im Osten an einer Küste, die felsig und voller Macchia ist, dornigem Gestrüpp, trockenen Wäldern und kargen Felsen – selbst die Buchten und Strände sind mit grobem Kies gefüllt. Wir saugen die Sonne auf, genießen den stetigen Wind und essen gegrillten Fisch – wie sehr wir das Meer vermisst haben, viele Tausende Kilometer weit, wird uns erst jetzt klar. Den Weg ins Landesinnere, zu den Plitvicer Seen nehmen wir trotzdem – und als wir dort sind, zwischen den weiß gischtenden Wasserfällen, türkisgrün leuchtenden Wasserbecken und tiefgrünen Seen in einem üppig grünen Tal, fühlen wir uns tatsächlich ein klein wenig an Südostasien erinnert. An grüne, feuchte Berge, die in der Hitze des Tages mit ein wenig Kühle locken. Heimweh? – Natürlich nicht. Nur eben etwas Erfrischung fürs Gemüt tanken.

Die Schönheit dieses Ortes fährt auf den nächsten Etappen-Kilometern jedenfalls mit. Zurück ans Mittelmeer, für einen endgültigen Abschied. Denn nun haben wir sie endgültig erreicht: die Alpen. An ihrem südöstlichen Ende landet das große Gebirge genau hier, wo sich Kroatien, Slowenien und der äußerste Zipfel Italiens am Golf von Triest für einen sommerlichen Moment treffen. Für unsere Porsche und uns ist das ein ganz besonderer Moment. Hier beginnt eine Welt, an deren anderem Ende unser

It's not far from here to Croatia and the Mediterranean and moments later we find ourselves on the rugged cliffs that take us on a journey of well over 500 kilometers to the north. This part of the Mediterranean is known as the Adriatic. On the other side lies Italy and here in the east we follow a rocky coastline full of maquis, thorny scrub, tinder-dry forests and barren crags – even the bays and beaches are filled with coarse gravel. We soak up the sun, enjoy the constant breeze and dine on grilled fish, only now realizing just how much we have missed the sea on this journey of many thousands of kilometers.

mountains and cool river valleys make an exception, allowing the birds to chirp, and lizards to crawl languidly over old stone walls. Life passes with wonderous ease. Beneath the »Stari Most«, the old bridge at Mostar, the Neretva River is an ice-cold swirl of green. We gaze down into the water and savor the carefree weightlessness of the day.

It's not far from here to Croatia and the Mediterranean and moments later we find ourselves on the rugged cliffs that take us on a journey of well over 500 kilometers to the north. This part of the Mediterranean is known as the Adriatic. On the other side lies Italy and here in the east we follow a rocky coastline full of maquis, thorny scrub, tinder-dry forests and barren crags – even the bays and beaches are filled with coarse gravel. We soak up the sun, enjoy the constant breeze and dine on grilled fish, only now realizing just how much we have missed the sea on this journey of many thousands of kilometers.Nonetheless, we cut inland to the Plitvice Lakes. There, between the white spray of the waterfalls, radiant turquoise-green basins and deep green lakes in a lush green valley landscape, we are actually a little reminded of Southeast Asia with its moist, verdant mountains that promise a little coolness in the heat of the day. Homesickness maybe? – Not for a moment. We are simply filling up on some refreshment for the mind. In any case, the beauty of this place will be with us for the next few kilometers of our journey back to the Mediterranean, for a final farewell. We have now finally reached the Alps.

Ziel liegt: die Porsche-Hauptstadt Stuttgart, nur eine oder zwei Autostunden vom Nordrand der Alpen entfernt. Und in Villach, hinter dem ersten Alpenkamm, beginnt das Porsche-Universum. Der Firmengründer hat wesentliche Jahre im kleinen Gmünd verbracht und gearbeitet, dorthin zieht es uns jetzt magisch: zuerst von Kroatien nach Slowenien, dann an der slowenischen Hauptstadt Ljubljana vorbei, und von hier im Autobahntunnel unter dem Karawanken-Gebirge nach Österreich – fünfzig Kilometer hinter Villach haben wir Gmünd erreicht. Ankommen in Etappen, nennt man das wohl ...

Nächster Stopp? – Die Großglockner-Hochalpenstraße. Die ersten Porsche-Auto-Generationen haben allesamt hier oben das Laufen gelernt. Jetzt fahren wir über den mautpflichtigen Pass mit seinen unzähligen Kehren. Schrauben uns aus dem sommerlich grünen Tal in die felsigen Höhenlagen jenseits der Baumgrenze hinauf. Holpern über Kopfsteinpflaster, tauchen in die dunklen Tunnel unter Felsgraten, und landen schließlich auf dem Parkplatz beim F.A.T. Mankeiwirt. Die Helden des Porsche Club Pinzgau sind uns bis hierher in die Höhe entgegengefahren. Genau jetzt beginnt es zu regnen – gut, dass wir uns in der uneingeschränkten Bewunderung der Porsche-Locals sonnen können. Schon klar, wir wissen ziemlich genau was hinter uns liegt, was wir geleistet haben, aber etwas Anerkennung tut dann irgendwie doch gut. Und während wir so hier oben sitzen, während mächtige Wolkenzüge steingrau gegen die Alpengipfel branden und ihre Wasserlast abwerfen, wird uns klar, dass wir uns jetzt auf der Zielgerade befinden: Selbst mit dem geplanten Umweg zum Treffen der Munich Gentleman Drivers am Salzburgring liegen nun nur noch 550 Kilometer vor uns – ein Klacks, verglichen mit den Etappen in Asien.

Wie beobachten uns dabei, wie wir tatsächlich versuchen, noch etwas Zeit zu schinden. Das Ende dieser Fahrt wird wie das Aufwachen aus einem bunten, lauten, herrlichen Traum sein. Verschwitzt und durcheinander, aber voller neuer Ideen und Flausen. Über die Einladung bei F.A. Porsche Design unten in Zell am See, gleich am Ende der nun doch wieder sonnig-trockenen Großglockner-Straße freuen wir uns daher ausgelassen. Wir nehmen alles mit, kosten die Momente nun noch viel intensiver aus. Baden im Gemeinschaftserlebnis am Salzburgring, fahren dann gespannt und beinahe andächtig auf die Autobahn in Deutschland. – Zum ersten Mal seit vielen Wochen klopft die Tachonadel zart an Geschwindigkeiten jenseits der 170, 180 km/h an. Ein paar ganz Wilde testen noch härtere Gangarten aus, sind

The southeastern end of the great mountain range lands right here where Croatia, Slovenia and the extreme tip of Italy meet on the Gulf of Trieste for a moment of summer sun.

This is a very special moment for ourselves and our Porsches, marking the beginning of a world that contains our final destination: the Porsche capital of Stuttgart is only an hour or two away by car from the northern edge of the Alps. Villach, beyond the first Alpine ridge, is where the Porsche universe begins. The company's founder spent a significant amount of time working in the small town of Gmünd, where we are now drawn as if by magic: we move from Croatia to Slovenia, then past the Slovenian capital Ljubljana, and from here we take the motorway tunnel under the Karawanken Mountains to Austria – fifty kilometers beyond Villach we reach Gmünd in what you might call a gradual arrival...

Our next stop is the Grossglockner High Alpine Road. The early generations of Porsche cars all made their maiden voyages up here. We now negotiate the tolled road with its countless hairpin bends, spiraling up from the summery green valley to the rocky heights above the tree line. We rattle over cobblestones, dive into the dark tunnels under rocky ridges, and finally end up in the parking lot at F.A.T. Mankeiwirt. The heroes of the Pinzgau Porsche Club have driven up to meet us here. Right on cue, it starts to rain, so it's good to bask in the unlimited admiration of the Porsche locals. Of course we're well aware of the distance we have covered and what we have achieved, but a pat on the back always feels good. As we sit up here, while mighty stone-gray clouds crash against the Alpine peaks and shed their watery load, we realize that we are now on the home stretch: even with a planned detour to meet with the Munich Gentleman Drivers at the Salzburgring, we now have only 550 kilometers ahead of us – a piece of cake, compared to the stages completed in Asia. We find ourselves actually trying to buy some more time. The end of this journey will be like waking up from a colorful, loud and wonderful dream. Sweaty and confused, but with heads full of crazy new ideas, we are happy to accept an invitation from F.A. Porsche Design down in Zell am See, right at the end of the now sunny and dry Großglockner Road. We take it all in, savoring the moments even more intensively. We wallow in the community experience at the Salzburgring, then drive excitedly and almost reverently onto the motorway in Germany. For the first time in many weeks, the speedometer needle delicately hits above beyond 170, 180 km/h. A few of our really wild companions push for

aber schnell wieder zurück in der Meute: Wir haben diese Reise gemeinsam begonnen und werden sie auch gemeinsam zu Ende bringen. Vorbei an München, Augsburg und Ulm, unterwegs treffen wir sogar noch ein letztes Mal auf die Donau. Graubraun schiebt der Fluss dahin, als deutlich kleinere Version seiner späteren Erscheinung. »Wenn du jetzt wüsstest, wie weit dein Weg noch ist,« denken wir in den ein, zwei Sekunden, die es braucht, um die Donau auf einer Brücke zu überfliegen. Wir wissen es, wir haben die Welt gesehen, in die es für den Fluss hineingeht ... Etwas später ändert sich auch für uns wieder einmal das Land, die Autobahn sackt plötzlich in einem engen Tal viele Höhenmeter nach unten, landet auf dem Plateau vor Stuttgart.

Jeder Einzelne fährt jetzt ganz in seine Gedanken versunken, nimmt die Kilometer beinahe übersensitiv wahr. Abschiedsschmerz von der Reise macht sich breit, als süßsaures Weh. Die blauen Schilder der deutschen Autobahn haben eine unbestechliche Botschaft für dich: fast angekommen, beinahe vorbei. Zögerlich fährst du, ungläubig und irgendwo zwischen euphorisch und todtraurig. Du hast Stuttgart erreicht, rollst von der Bahn hinunter in den Talkessel. Vorbei am Mercedes-Stern auf der Firmenzentrale der anderen Stuttgarter Automobil-Legende, und dann quer durch die Stadt, bis zum Porsche Museum und dem Porsche-Headquarter in Stuttgart-Zuffenhausen. 60 Tage, 17 Länder und 22.000 Kilometer liegen hinter dir, völlig verrückte Zahlen, die dich ungläubig den Kopf schütteln lassen.

Das Ende der Fahrt ist bemerkenswert: Sie hört einfach auf. Du rollst im Konvoi der Freunde zum Finale und plötzlich taucht eine kleine Zielflagge auf dem Display des Navigationsgeräts auf. Die letzten Kilometer werden heruntergezählt, und wenn es dann nur noch Meter sind, würdest du am liebsten umdrehen und zurückfahren. Oder weiter. Alles, nur nicht anhalten. Denn das kommt dir nach den letzten 22.000 Kilometern vollkommen unnatürlich vor. Aber du wirst erwartet, von Freunden und Familie. Der Stuttgarter HEIZR Club hat uns eingeladen, der Porsche Brand Store und die Museums Crew möchten uns sehen und das klingt irgendwie spannend. Neue Abenteuer voraus, von den vergangenen erzählen und dabei die vielen zurückliegenden Wochen verarbeiten. Das Auto rollt aus, du lauschst ein letztes Mal dem selbstbewussten Lauf des Motors. Dann: Zündschlüsseldreh. Aus. Draußen siehst du lachende Menschen, das Begrüßungskomitee ist aus dem Häuschen. Einen kurzen Moment atmest du noch ein. Atmest aus. Und dann öffnest du die Fahrertür.

even more speed but quickly rejoin the pack: we started this journey together and intend finishing it together. We bypass Munich, Augsburg and Ulm, even coming face-to-face with the Danube for one last time along the way. A much diminished version of the waterway we encountered earlier, the gray-brown waters of the river roll onward. »If you only knew now just how far you still have to go,« we think as we cross the Danube by bridge in a matter of seconds. We know what we're talking about. After all, we have seen the world the river is about to enter... A little later, the countryside changes for us yet again, as the highway suddenly drops dramatically into a narrow valley, landing on the plateau just ahead of Stuttgart.

Each of us is now completely lost in our thoughts as we drive, experiencing the last few kilometers with something bordering on hypersensitivity. The pain of saying goodbye to the journey wells up inside us as a sweet sorrow. The blue signs of the German autobahn have an undeniable message for us: you are almost there; the adventure is almost over. Incredulous, you drive hesitantly, hovering somewhere between euphoria and desperate sadness. You reach Stuttgart, descending from the highway down into the valley, past the Mercedes star at the headquarters of Stuttgart's other automotive legend, and then across the city to the Porsche Museum and the Porsche headquarters in Stuttgart-Zuffenhausen. 60 days, 17 countries and 22,000 kilometers lie behind you, completely crazy statistics that make you shake your head in disbelief.

The end of the journey is quite remarkable for its suddenness. You cruise to the finish in a convoy of friends and suddenly a small checkered flag appears on the display of the navigation system. The last few kilometers are counted down, and when it's only a matter of meters, you feel like you'd gladly turn around and drive back again. Or maybe keep going. Anything but stopping, which seems completely unnatural to you after the last 22,000 kilometers. But friends and family await. The Stuttgart HEIZR Club has invited us to pay a visit, while the Porsche Brand Store and the museum crew also want to see us, all of which sounds kind of exciting. You think of new adventures ahead and the tales of past experiences, trying to process what has happened over the last few weeks. The car comes to a stop, and you listen to the self-confident sound of the engine one last time. Then you turn off the ignition. Outside you see happy, smiling faces. The welcoming committee is over the moon. You take a short breath, then exhale. Finally, you open the driver's door.

RESTAURANT

AA 161
THAILAND

THE EPIC BANGKOK STUTTGART PORSCHE DRIVE

Nach einem Monat Pause treffen wir uns in Tiflis. Wir begegnen einer befreundeten Motorradgruppe aus Thailand, die wir in den nächsten Tagen immer wieder treffen. Unsere Truppe ist etwas größer geworden – einige Freunde begleiten uns nun auf dem Weg. After a one-month break, we meet again in Tbilisi. We run into a friendly motorcycle group from Thailand whom we keep meeting again over the following days. Our group has grown a bit larger – a few more friends are now accompanying us on the journey.

Porsche Club
DRI. CHAYANIN D.
CO.DRI. GON N.
TAG HEUER
XPEL
THE EPIC BANGKOK STUTTGART PORSCHE DRIVE
Plan-B
TROS

CHAYANIN
Reinheitsgebot
Leica
PORSCHE

50
CURVISTAN
LHD
TOW
4C 66
THAILAND 01
SINGHA

20

LUKMED

SINGHA

RENNDRIVE
66

Das ist genau das, was du suchst: Gelände. Die Federbeine von »Offroad Cayman«, »Safari« und »Dakar« frohlocken – die Bandscheibe meldet sich, aber die Freude ist groß. Georgien ist rau und massiv. We like it that way! Yeah! This is exactly what you're looking for: terrain. The suspension of »Offroad Cayman,« »Safari,« and »Dakar« is celebrating – the spine, not so much, but the joy is huge. Georgia is raw and massive. We like it that way! Yeah!

Hier wären wir liebend gern noch weiter durch die Landschaft gefahren, aber unseren 992 GTS hätten wir zurücklassen müssen. Next time …
We would have loved to keep driving through the landscape, but we would have had to leave our 992 GTS behind. Next time…

BO 992
THAILAND

Willkommen in der Türkei: Kurz mal am Strand festgefahren und den obligatorischen Bremsencheck am Straßenrand erledigt. Welcome to Türkiye: Got briefly stuck on the beach and did the obligatory brake check at the roadside.

Leica
STUTTGART

DRI. CHAYANIN D.
CO.DRI. GON N.
Porsche Club
TAG HEUER
NGHA
MON SODA
THE EPIC BANGKOK STUTTGART PORSCHE DRIVE
XPEL
Plan-B
S TREFFEN
TROS

THE EPIC BANGKOK STUTTGART PORSCHE DRIVE

Die Türkei ist ein Traum zum Fahren – und zum Essen. Die Landschaft ist fantastisch. Wir als Truppe haben uns wieder bestens eingegroovt – jeder Tag ein Höllenspaß. *Turkey is a dream for driving – and for food. The landscapes are fantastic. As a group, we've found our rhythm again – every day is an absolute blast.*

Wie haben die das hier hochgebracht? Das berühmte Monument mit Grabmal auf einem Berg in der Türkei ist der Nemrut Dagi, auf dessen Gipfel König Antiochos I. von Kommagene ein monumentales Heiligtum errichten ließ. Das Grabmal besteht aus einem über 50 Meter hohen Grabhügel aus Geröll, umgeben von drei Terrassen mit riesigen – heute größtenteils kopflosen – Statuen griechisch-armenischer und persischer Götter sowie des Königs selbst. How did they get all of this up here? The famous monument with a tomb on a mountain in Türkiye is Mount Nemrut, where King Antiochus I of Commagene had a monumental sanctuary built at the summit. The tomb consists of a gravel mound over 50 meters high, surrounded by three terraces with enormous – now mostly headless – statues of Greco-Armenian and Persian gods, as well as the king himself.

Level 56.
loading..., begins!
Happy Birthday

HCG TRAVEL GROUP
华程国旅集团
Corendon

Kappadokien ist für seine markanten Tuffsteinformationen, historischen Höhlenkirchen und weitläufigen unterirdischen Städte bekannt. Ebenso berühmt sind die Heißluftballonfahrten, bei denen bei Sonnenaufgang dutzende Ballons über die Täler aufsteigen und die Landschaft aus der Luft sichtbar machen – etwas touristisch, aber absolut sehenswert. Cappadocia is known for its distinctive tuff rock formations, historic cave churches, and extensive under-ground cities. Equally famous are the hot-air balloon rides, where dozens of balloons rise over the valleys at sunrise and reveal the landscape from above – a bit touristy, but absolutely worth it.

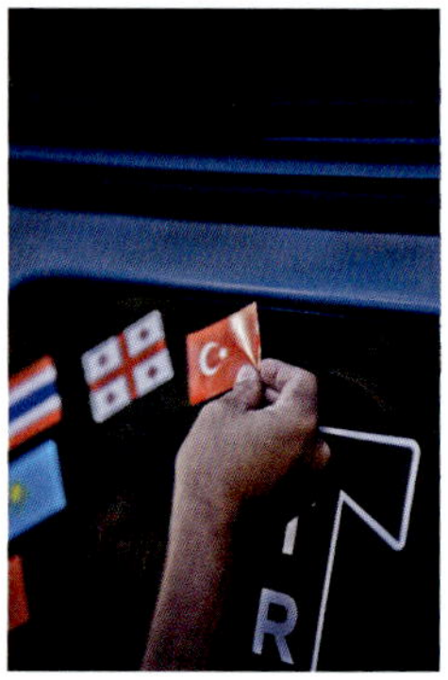

Flower
Balloons

OSFER

SET
東南旅行社
www.settour.com.tw
鳳凰旅遊
PHOENIX TOURS
KALSEDON
A533807
THAILAND

Die Bibliothek von Ephesos, auch Celsus-Bibliothek genannt, wurde im 2. Jahrhundert n. Chr. errichtet und war mit rund 12.000 Schriftrollen eine der größten Bibliotheken der römischen Antike. Besonders berühmt ist ihre reich verzierte, heute teilweise rekonstruierte Fassade, die zu den eindrucksvollsten Bauwerken der antiken Welt zählt. *The Library of Ephesus, also known as the Library of Celsus, was built in the 2nd century AD and, with around 12,000 scrolls, was one of the largest libraries of the Roman world. It is especially famous for its richly decorated, partially reconstructed façade, which is considered one of the most impressive structures of the ancient world.*

ROYCE
WELCOME
RIPE
TOMATOES

LAND-
ROVER
PORSCHE
ROLLS
ROYCE

HARLEY-
DAVIDSON

1915ÇAN
20
SINGHA
LEMON SODA
DAS TREFFEN
Leica
4C 66
THAILAND

GELİBOLU GÜNEY GİŞELERİ
Geniş Araç
HGS
NAKİT
KKARTI
Gelibolu
HAREKET ETMEK
TEHLİKELİ VE YASAKTIR
10m
20
10m
20

CURVISTAN
BO 992
THAILAND

BO 992
THAILAND

Das konnten wir uns nicht entgehen lassen. Statt des Dracula-Schlosses ging es für uns zum Buzludscha-Monument auf einem Berggipfel im bulgarischen Balkangebirge – eines der ikonischsten Beispiele für sozialistischen Brutalismus in Europa. Der UFO-artige Betonbau wurde 1981 als Gedenkstätte der Kommunistischen Partei eröffnet und ist heute verlassen, aber gerade dadurch beeindruckend.

We couldn't pass this up. Instead of visiting Dracula's Castle, we headed to the Buzludzha Monument on a mountain peak in Bulgaria's Balkan Range – one of the most iconic examples of socialist brutalism in Europe. The UFO-like concrete structure was inaugurated in 1981 as a memorial to the Communist Party and is now abandoned, which makes it all the more impressive.

BO 992

CURVISTAN
4C 66
THAILAND

Wir sind schon viele Pässe gefahren – doch dieser hier fühlt sich an, als wäre er für Fahrer gemacht. Der Transfagarasan windet sich wie eine kleine Carrera-Bahn durch das Gebirge, perfekt in die Landschaft gelegt und mit jeder Kurve ein neues Highlight. We've driven many mountain passes – but this one truly feels purpose-built for drivers. The Transfagarasan twists through the mountains like a miniature slot-car track, perfectly carved into the landscape and offering a new highlight with every turn.

RENNDRIVE
66

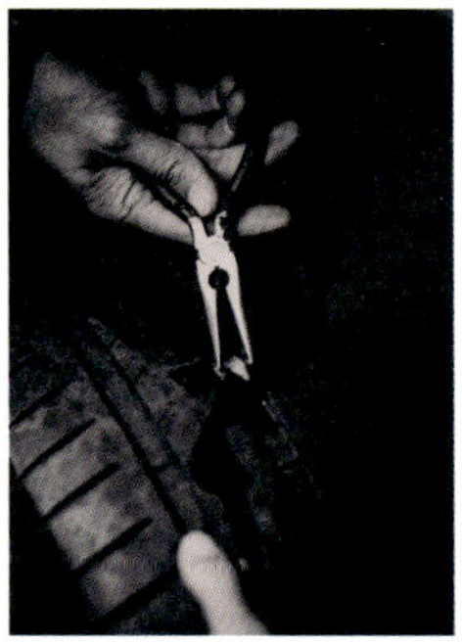

PEST-BUDA

Unser 992 GTS hat auf den letzten 500 Kilometern immer wieder Luft verloren – wie sich herausstellte, lag die Ursache in einem Felgenriss. Eine Ersatzfelge war so schnell nicht zu bekommen, also hieß es: ran ans Schweißgerät! Our 992 GTS kept losing air over the last 500 kilometers – and as it turned out, the cause was a cracked rim. A replacement wheel wasn't available on short notice, so it was time to bring out the welding equipment!

A53 3807
THAILAND

PORSCHE

CURVISTAN
BO 992

SREBRENICA
1995 - 2025

BO 992
THAILAND

Endlich wieder am Meer – und dazu gleich eine absolute Schmankerlstraße hinunter zur Küste. Finally back at the sea – and right away a real gem of a road leading down to the coast.

66

ANTIQUE

STUTTGART
BANGKOK
THE EPIC BANGKOK STUTTGART PORSCHE DRIVE

Die Plitvicer Seen sind ein Muss für jeden Naturliebhaber und Wanderfreund. Eine traumhafte Gegend und ein wunderbarer Ausgleich abseits des Schalensitzes. The Plitvice Lakes are a must for every nature lover and hiking enthusiast. A stunning region and a wonderful change of pace away from the bucket seat.

Nacionalni park
National park
GALOVAC
KOZJAK

SINGHA

THE EPIC BANGKOK STUTTGART PORSCHE DRIVE
4C 66 THAILAND

Glockenhütte

Wiedweg

GLOCKENHÜTTE

Für uns ein wichtiges Ziel auf unserer Pilgerfahrt: Das Porsche Automuseum Pfeifhofer in Gmünd ist das weltweit erste private Porsche-Museum und widmet sich besonders den frühen Jahren der Marke. Da in Gmünd die ersten 356-Modelle entstanden, zeigt das Museum seltene Originale und macht die Anfänge der Porsche-Geschichte unmittelbar erlebbar.
An important stop on our pilgrimage: The Porsche Automuseum Pfeifhofer in Gmünd is the world's first private Porsche museum and focuses especially on the brand's early years. Since the first 356 models were built in Gmünd, the museum displays rare originals and brings the beginnings of Porsche history to life.

PORSCHE AUTOMUSEUM
HELMUT PFEIFHOFER

Kehre 6

CURVISTAN
BO 992
Ende

BO 992
THAILAND

BANGKOK
STUTTGART
BO 992
THAILAND

RS

50

Jetzt geht's Schlag auf Schlag: Auf dem Großglockner werden wir im F.A.T. Mankei vom Porsche Club Pinzgau empfangen, und in Zell am See gibt es eine Privatführung durch das F.A. Porsche Design Studio. Weiter geht's dann ins Hans-Peter-Porsche-Museum – Druckbetankung vom Feinsten. Now things are happening one after another: On the Grossglockner, we were welcomed at F.A.T. Mankei by the Porsche Club Pinzgau, and in Zell am See we enjoyed a private tour of the F.A. Porsche Design Studio. After that, we're heading to the Hans-Peter Porsche Museum – full-throttle immersion at its finest.

FUSCHER LACKE
PARKPLATZ
2262 m ü.d. M.

GREETINGS FROM GOODWOOD
GREETINGS FROM GP ICE RACE

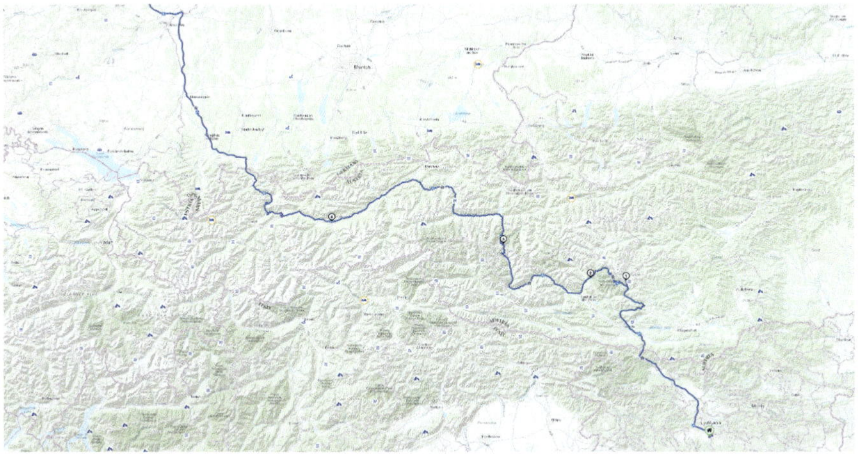

Porsche Club
Pinzgau
welcome our Porsche friends
from Thailand

RS

BANGKOK
STUTTGART
RENNDRIVE
BO 992
THAILAND

STUDIO F.A.PORSCHE

ENERGY STATION
DM
Liter
bleifrei
BOX BOX BOX
SUPER BREW
BOX BOX BOX
Rauchen verboten
Motor u. Fremdheizung abstellen

ZIMMER FREI

Seewirt

Gelateria San Marco
Restaurant · Café · Derbl
Heritage.Hotel
Restaurant.Café

Kurzer Boxenstopp am Salzburgring zum Besuch bei den Munich Gentleman Drivers. A quick pit stop at the Salzburgring to visit the Munich Gentleman Drivers.

6

POLIZEI
POLIZEI

CURVISTAN
AAS
TAG HEUER
BO 992
THAILAND 01
AAS

Am Ziel: Nach 60 Tagen, 17 Ländern und 22.000 Kilometern sind wir ohne große Komplikationen in Stuttgart angekommen. Danke an die HEIZR-Crew und den Porsche Brand Store für die Willkommenseinladungen. Der obligatorische Porsche-Shot am Museum durfte natürlich auch nicht fehlen – danke auch an die Museums-Crew! At the finish line: After 60 days, 17 coun-tries, and 22,000 kilometers, we arrived in Stuttgart without major complications. Thanks to the HEIZR crew and the Porsche Brand Store for the warm welcome. The obligatory Porsche shot at the museum was, of course, a must – and thanks as well to the museum crew!

LH60
industry
KAATSCH
LIEBHERR
19
S-ZL 229

Team-Preis
8 PORSCHE AM START

CIBIE
Shell

TENN – THANK YOU:

This journey, from the bustling heart of Bangkok to the iconic streets of Stuttgart, has been nothing short of an odyssey – a profound, life-changing experience etched forever in our souls.

To every incredible friend who shared a piece of this adventure, who lent their spirit, their time, and their unwavering support, my deepest gratitude. This dream would have remained just that, a dream, without you.

»Khun A«: the visionary spirit who first ignited this audacious dream and tirelessly pushed us to make it a reality. His unparalleled generosity knew no bounds. When a setback threatened our plans, he, without hesitation, offered his own magnificent 992 GTS Thailand edition (one of only thirty ever made!) to join our convoy all the way to Stuttgart. His kindness is a beacon.

»Chayanin« and his extraordinary team – P Mog and Joke: Chayanin, the brave soul who piloted his 911 Carrera Rally car – a symphony of raw power with no carpet, a full cage, and oil cooler lines running through, truly a »flying oven«. And P Mog and Joke, his tireless guardian angels, formed the backbone of our expedition. Their indispensable Toyota Land Cruiser service vehicle, laden with spare parts for Chayanin's Carrera and our own vital spare wheels, offered us an invaluable peace of mind, knowing that whatever went wrong, they had our backs. We truly could not have made it without them.

»Oou«: my steadfast partner in crime, whose passion shone brightly as he meticulously restored his 987 Cayman S, breathing new life into it specifically for this monumental trip!

»Tee«: my other intrepid co-conspirator. Our initial plans to complete our 996 C4s hit a roadblock, but the universe, in its infinite wisdom, sent **Suthi,** who, with an open heart, stepped in and offered his own 992 GTS, ensuring our dream remained on course.

»Porsche Thailand and AAS Motorsport«: for their invaluable support, providing the trusty brand new Cayenne S E-Hybrid. Assembled with care in Malaysia especially for the Thai market, this magnificent vehicle became the unwavering backbone of our convoy, carrying extra passengers and crucial gear, mile after tireless mile.

»Maj Jack«: and his rugged G-Wagon, was another crucial pillar of our group. His G-Wagon, more than just a vehicle, was our mobile command center, laden not only with essential gear but also vital first-aid supplies. His military precision and calm demeanor, born from his background, were a constant source of reassurance throughout the journey.

»A-GT and Art Racing«: our silent heroes behind the wheel – their willingness to step in and offer relief driving was a gift, ensuring our spirits remained high and our journey safe, allowing weary drivers to rest.

»Ake« our seasoned guide, whose intimate knowledge of these very roads transformed a mere path into a well-trodden, confident route, making the impossible feel navigable.

»Stefan Bogner«: our friend and talented photographer. He didn't just document our journey; he poured his heart and soul into capturing every precious moment, crafting this magnificent book for us. His relentless energy, paired with a truly keen eye for those breathtaking, epic shots, has given us an enduring gift – a timeless treasure that ensures the spirit of this trip will live in our memories forever.

We also extend our immense gratitude to our Renndrive friends, who joined us along the way and celebrated with us at the end – your camaraderie was a true joy.

But this adventure was also woven with countless threads of human connection. The many friends we met along the way, who made us feel so incredibly welcome and safe. The Porsche clubs that greeted us with open arms, taking us for exhilarating drives – the »Porsche camaraderie« isn't just a phrase; it's a living, breathing force! Some truly saved our behinds, like **Miki,** who, without hesitation, took us to fix a cracked wheel, ensuring we could march on.

So, thank you, from the deepest part of my heart, to every single soul who touched this epic journey. This wasn't just a drive; it was a tapestry of shared dreams, unwavering support, and unforgettable moments that will resonate within us for the rest of our lives. Every person we met has left an indelible footprint in our lives, forever changing us for the better.

STEFAN – THANK YOU:

I met the »Renndrive Crew« at *Das Treffen 3* in Bangkok – and since then, we have met countless times all over the world and experienced many small adventures together. Thai hospitality is truly unique in the world. I would like to thank you for the friendship and the trust, and for taking me along on this journey.

A very special thanks goes to K. Chayanin – without him, I would not have been able to embark on this journey. Without him, this book would not exist.

Many thanks to the people at Porsche in Zuffenhausen, Bangkok and Singapore, who have supported me over the years with a wide variety of projects in such an uncomplicated way. You are something like a power bank for me.

Thank you to the crew at the Porsche Museum and the Brand Store, who made our arrival a very special event.

Thanks to everyone at Porsche Design, especially the design team in Zell am See.

Thank you Ferdi & F.A.T Mankei for the great welcoming at Grosglockner.

Thanks to all the Porsche Clubs who welcomed us in their countries along the long journey. Thanks as well to the Heizr Crew for the Heizr Day.

Thanks to my family, my wife and my two sons, for giving me the freedom to drive the world.

A special thank-you goes to »Ake«, who planned the tour and guided us like a North Star. I have never experienced a journey that ran so smoothly and effortlessly.

Thank you Ben for beeing the invisible traveller along the trip.

And finally: Suthi, for the absurd idea of this epic trip.

THANK-YOU: ALL PORSCHE FANS WHO HAVE BEEN ACCOMPANYING AND VISITING US FOR ALMOST TWO DECADES NOW, ENCOURAGING US AND WELCOMING US EVERYWHERE.

THANK YOU FOR YOUR KIND SUPPORT:
SINGHA & PORSCHE LIFESTYLE

SINGHA
—LEMON SODA—

PORSCHE DESIGN

KONZEPT/CONCEPT:
Stefan Bogner,
Curves Magazin

LAYOUT:
Stefan Bogner

TEXT:
Ben Winter

ÜBERSETZUNG/TRANSLATION:
Jim O'Neill

LEKTORAT/EDITOR:
Hanno Vienken

FOTOS/PHOTOS:
Stefan Bogner

MOTIVAUSARBEITUNG/ PHOTO PROCESSING, ARTWORK, LITHOGRAFIE/LITHOGRAPHY
Michael Dorn

DRUCK/PRINTED BY:
Druckerei Vogl GmbH & Co. KG

KARTENABBILDUNG S. 4/5
www.istockphoto

KARTEN
www.kurviger.com

PRINTED IN GERMANY 2026

Bibliografische Information der Deutschen Nationalbibliothek
Die Deutsche Nationalbibliothek verzeichnet diese Publikation in der Deutschen Nationalbibliografie; detaillierte bibliografische Daten sind im Internet über http://dnb.dnb.de abrufbar.

Bibliographic information published by the Deutsche Nationalbibliothek

The Deutsche Nationalbibliothek lists this publication in the Deutsche Nationalbibliografie; detailed bibliographic data are available on the Internet at http://dnb.dnb.de.

1. Auflage / 1st edition
ISBN 978-3-667-13213-0

Delius Klasing Verlag GmbH
Siekerwall 21, D-33602 Bielefeld, Germany
Telefon/Phone +49 (0)521 559-0
Email: info@delius-klasing.de
www.delius-klasing.de

This book is dedicated to the
people who supported this trip.

Especially to K. Chayanin & Singha,
and the believers from Porsche &
Porsche Design.

ENGLANDS SCHÖNSTE GARTENSCHÄTZE

CALLWEY